ELEMENTS OF NUCLEAR REACTOR ENGINEERING

ELEMENTS OF NUCLEAR REACTOR ENGINEERING

L. Wang Lau.

Tennessee Valley Authority

GORDON AND BREACH SCIENCE PUBLISHERS

London   New York   Paris

Copyright © 1974 *by*

Gordon and Breach, Science Publishers, Inc.
One Park Avenue
New York, N.Y. 10016

Editorial office for the United Kingdom

Gordon and Breach, Science Publishers Ltd.
42 William IV Street
London W.C.2.

Editorial office for France

Gordon & Breach
79 rue Emile Dubois
Paris 75014

034 9920 2

D
621.483
LAU

To FLORENCE
Rita
Serena

PREFACE

Sir Isaac Newton once said: "If I have seen further than other men, it is because I have stood on the shoulders of giants." When an author like Newton, whose work is original, pays such a great tribute to his heritage, how humble and how much greater must be the debt of an author whose book contains no original work because of its very nature.

This book is designed to be used either as a textbook for a formal course in introductory nuclear engineering or as a useful supplement to all current texts. It is written to be used in one semester and the prerequisite is one year of university physics and one semester of sophomore differential equation. The text is primarily for undergraduates and graduates from other fields who want to find out in one semester of study what nuclear engineering is generally about so that they may know how and where it may be of use to them in their particular specialty, or that they may have the general background to go immediately into more classified advanced courses in the field of nuclear engineering. The book is also intended for nuclear engineering majors who are looking for a text with numerous illustrative problems so that their understanding of the basic principles and concepts of the subject can be best accomplished. One of the main objectives of the book is to bring the student closer to the realization of how various science and engineering fields are related.

As vast a field as nuclear engineering is, and as limited in size as a one-semester text has to be, it is natural that many topics have to be simplified. However, no effort has been spared in emphasizing completely solved problems which are meant to be an integral part of the text. The book is written in a very concise form and all detailed derivations are included in the section of solved problems so as to maintain an uninterrupted train of thought.

The supplementary problems are designed to cover all the material in each chapter and are not too numerous so that the student may attempt to work all of them.

It is recognized that nuclear constants and other data are continually being updated as a result of new experimental results. However, no serious attempt was made in getting the most up-to-date data for this introductory text since it is felt that the data given here are sufficiently accurate to allow for "quick-and-dirty" calculations to be made and to provide a basic general understanding of the subject at the beginning level.

The author wishes to thank Dr. Paulinus Shieh of Mississippi State University for reviewing the manuscript and for many helpful suggestions. To numerous authors, publishers, and companies, the author is indebted for their kind permission to use various data and figures. It is with gratitude that the author acknowledges the encouragement that his family has given him.

L.W.L

CONTENTS

Preface vii

1. NUCLEAR REACTIONS 1

   A. Classifications 1
      1. Scattering (Elastic Scattering; Inelastic Scattering) 1
      2. Absorption (Radioactive Decay; Radiative Capture; Fission; Transmutation) 1
      3. Spallation 2
      4. Fusion 2
   B. Threshold Energy and Q Energy 2
   C. Semiempirical Mass Formula and Binding Energy . 3
   D. Radioactive Decay 4

2. NEUTRON SOURCES AND DETECTION . . . 11
   A. Definitions 11
      1. Maxwellian Distribution 11
      2. Thermal Neutron 11
      3. Cross-section 12
      4. Flux 12

   B. Neutron Sources 13
      1. (α, n) Reaction with Light Elements 13
      2. (γ, n) Reaction with Be and H^2 13
      3. (d, n) and (p, n) Reactions with Light Elements 13
      4. Stripping of Deuterons 15
      5. High Energy (p, n) Reaction 15
      6. Fission Neutrons from a Nuclear Reactor 15
      7. Thermal Neutrons by Slowing-down 15
      8. Californium 252 15

   C. Neutron Detection and Energy Determination . . 15
      1. BF_3 Counter 15
      2. Long Counter 16

3.   Time-of-Flight Method and Neutron Chopper   16
4.   Neutron Crystal Spectrometer   17
5.   Activation Analysis   17
6.   Threshold Detector   18
7.   Fission Chamber   19
8.   Proton Recoil Counter   19
9.   Chemical Detectors   19
10. Others   19
11. Compensated Ion Chamber   20

3. FISSION   23

A.   Fissionable Materials—General Statements   .   .   23
B.   Fission Energy   23
C.   Experimental Data   24
D.   Prompt Neutrons   24
E.   Delayed Neutrons   32
F.   Prompt Gammas   34
G.   Delayed Gammas and Betas   34
H.   Fission Cross-section and Non-1/v Factor .   .   .   36

4. NUCLEAR REACTOR PLANT—
    GENERAL CONSIDERATIONS   41

A.   Nuclear Engineering   41
B.   Reactor Power Plant   41
C.   Specifications of Nuclear Reactors   42
D.   Reactor Analysis   43
E.   Reactor Characteristics and Design Data   .   .   .   45
F.   Engineered Safety Features   45
G.   Plant Site Considerations   51
     1.   Population and Water Supply   51
     2.   Meteorology   52
     3.   Hydrology   52
     4.   Geology   52
     5.   Seismology   53
H.   Plant Design   53
     1.   Containment   53
     2.   Engineered Safety Features   54
          a.   Emergency Core Cooling System   54
          b.   Containment Isolation System   55

      c.   Containment Cooling System 55

      d.   Emergency Air Treatment System 56

      e.   Other ESF Systems 56

   3. Auxiliary Systems 56

5. NEUTRON CYCLE   57

   A.   Introduction   57

   B.   Neutron Slowing-down   57

   C.   Resonance Absorption and Scattering   .   .   59

   D.   Neutron Diffusion and Leakage   .   .   .   60

   E.   Diffusion Equation   61

   F.   Age Equation   62

   G.   Thermal Utilization   63

   H.   Resonance Escape Probability   .   .   .   64

   I.   Fast Fission Factor   64

   J.   Infinite Multiplication Factor   .   .   .   65

   K.   Fast Non-leakage Probability   .   .   .   65

   L.   Thermal Non-leakage Probability   .   .   .   66

   M.   Effective Multiplication Factor and Criticality Condition   66

6. REACTOR FLUX DISTRIBUTIONS AND CRITICALITY
    CONDITIONS   73

   A.   Bare Homogeneous Core, One Group Method   .   .   73

   B.   Homogeneous Core with Reflector, One Group Method   77

   C.   Homogeneous Core with Reflector, Two Group Method   80

   D.   Heterogeneous Cell   85

      1. Thermal Utilization 89

      2. Resonance Escape Probability 89

      3. Age and Diffusion Length 90

      4. Fast Fission Factor 91

      5. Criticality Condition 93

7. REACTOR KINETICS AND CONTROL   .   .   97

   A.   Reactor Kinetics   97

      1. Kinetic Equations 97

      2. Inhour Equation and Reactivity 98

      3. One-delayed-neutron-group Approximation 100

B. Control Rod Theory   102
    1. Fully Inserted Rod   102
    2. Partially Inserted Rod   104
C. Reactor Control—General Considerations .   .   .   105
    1. Temperature Effect   105
    2. Fission Product Poisoning   107
    3. Fuel Depletion   108

8. HEAT TRANSFER AND FLUID FLOW   .   .   .   111

    A. Heat Removal—General Considerations   .   .   .   111
    B. Conduction, Convection, and Radiation   .   .   .   111
    C. Determination of Heat Transfer Coefficient   .   .   113
    D. Temperature Distributions   114
        1. Axial Temperature Distribution of the Coolant   115
        2. Radial Temperature Distribution of the Fuel   115
        3. Axial Temperature Distribution of the Fuel   116
        4. Maximum Temperatures   117
    E. Boiling and Critical Heat Flux   119
    F. Fluid Flow   120
        1. Frictional Pressure Drop   120
        2. Abrupt Changes in Flow Areas   121
    G. Thermal Constants of Reactor Materials .   .   .   122

9. RADIATION HAZARD, PROTECTION GUIDES, AND
    DETECTION   133

    A. Definitions   133
        1. Standard Man   133
        2. Curie   134
        3. Roentgen   134
        4. Rad   134
        5. Relative Biological Effectiveness   134
        6. Rem   136
        7. Mass Absorption Coefficient and Energy Absorption
            Coefficient   136
    B. Biological Considerations   137
        1. Biological Variability   137
        2. Latent Period   137

3. Recovery and Time Factor   137
4. Radiosensitivity   137
5. Relative Biological Effectiveness   137
6. Genetic Effect   137

C. Occasional and Accidental Exposures   .   .   .   138
D. Protection Criteria   139

1. Accumulated External Dose   139
2. Emergency and Medical Dose   140
3. Accumulated Internal Dose   140
4. Unidentified Radionuclides in Air and Water   140
5. Flux to Dose Rate Conversion   143

E. Radiation Detection   143

1. Gas-filled Detectors   145

   Geiger Counter   145
   Ionization Chamber, Pocket Meters, and
   Cutie Pie Meter   146
   Proportional Counter   147
2. Scintillation Detector   147
3. Film Dosimeter   148
4. Others   148
5. Neutrons, Alphas, and Beta Detection   148

F. Statistics of Counting   149

10. SHIELDING   157

A. Radiation Sources   157

1. Gamma Sources   157
2. Neutron Sources   158

B. Interaction of Gammas with Matter   .   .   .   158

1. Photoelectric Effect   159
2. Compton Effect   160
3. Pair Production   164

C. Build-up Factor   164
D. The Monte Carlo Method (Random Sampling)   .   .   166
E. Relaxation Length and Removal Cross-section   .   .   168
F. Shielding Materials   169

1. Iron   169
2. Lead   169

3. Concrete   169
4. Water   170
5. Others   170

G. Geometric Considerations   170

1. Point Kernel   171
2. Line Source   172
3. Disk and Infinite Plane Source   173
4. Spherical Surface Source   175
5. Semi-Infinite Slab (Finite Thickness) Source   176
6. Solid Cylindrical Source   177
7. Solid Sphere Source   178

H. Rules of Thumb   178

11. FUEL CYCLE AND ECONOMICS   187

A. Mining   187
B. Ore Concentration   187
C. Concentrate Purification   189
D. Enrichment   190

1. Gaseous Diffusion   191
2. Gas Centrifuge Process   191
3. Separation Nozzle Method   192
4. Others   193

E. Ammonium Diuranate (Adu) Process   .   .   .   193
F. Uranium Dioxide   194
G. Fabrication   194
H. Burn-up   195
I. Reprocessing   196
J. Waste Disposal   196
K. Fuel Loading   197
L. Fuel Economics   198
M. Nuclear Power Plant Cost   199
N. Uranium Fuel Reserves   200

12. PROGRESS IN REACTOR ENGINEERING   .   .   201

A. The Nuclear Industry   201
B. Contemporary Uses of Nuclear Power   .   .   .   201

1. Space System   201
2. Plowshare—Peaceful Uses of Nuclear Explosives   205

3. Pollution   206

C. High Temperature Gas-cooled Reactor . . . 209
D. Liquid Metal Fast Breeder Reactor . . . 211
E. Controlled Fusion Power 213

13. CRITERIA, STANDARDS, AND GUIDES . . . 215

A. Definitions and Explanations 215
B. General Design Criteria 216
C. Safety Guides 217
D. Codes 219
E. Emergency Core Cooling System—Interim Policy
     Statement 220
F. ANS Safety Classes 221
G. Safety Analysis Report Format 221
H. ASME Section III 222
I. Electrical Safety Class 222

Appendix A. Miscellaneous Constants 223

         B. Conversion Table 225

         C. Greek Alphabet 229

Subject Index 231

Chapter 1

NUCLEAR REACTIONS

THE UNDERSTANDING of nuclear reactions is essential to the understanding of the production of nuclear power. The following sections provide a general background in this respect.

A. Classifications

Nuclear reactions can be caused by the bombardment of a nucleus with proton, neutron, gamma ray, deuteron, alpha particle, or other elementary particles and heavy nuclei. Of particular interest to nuclear engineers concerning these reactions are:

1. the types and products of the reaction
2. the energy required for or produced by the reaction, and
3. the probability (or cross-section) of occurrence.

There are many different ways of classifying the reactions, such as that by energy consideration or by the types of incident particles involved. A simple and useful method is to classify them by the types of reactions involved and the particles emitted:

1. *Scattering*—the incident particle is of the same kind as the outgoing particle.

a) Elastic scattering—the kinetic energy of the system is conserved, although there is a transfer of kinetic energy and momentum between the incident particle and the target nucleus. There is no change in the internal, or nuclear, state in the target nucleus.

b) Inelastic scattering—the target nucleus is transformed into an excited state and has higher potential energy. The total kinetic energy is not conserved in the process. The excited target nucleus usually decay very quickly to a lower energy state by emitting gamma rays.

2. *Absorption*—the incident particle or radiation is absorbed by the target nucleus, forming an unstable compound nucleus. Several processes may follow as a result of instability:

a) Radioactive decay—alphas or betas are emitted, usually accompanied by gamma rays. The decay rate is exponential in time.

1

B

b) Radiative capture—gamma rays are emitted almost instantaneously.
c) Fission—compound nucleus breaks up into two or more heavy frag-
    ments of lower mass number and several neutrons are ejected by these
    fragments. Most neutrons are emitted promptly (*prompt neutrons*) but a
    small fraction is delayed (*delayed neutrons*). The fragments carry a tre-
    mendous amount of kinetic energies and are often radioactive. Prompt
    gammas and neutrinos also accompany the fission process.
d) Transmutation—protons or heavier particles are emitted very quickly
    and the original target nucleus is transformed into a nucleus of a new
    different element.
3. *Spallation*—When the incident particle has energy of over 100 Mev, such
as that in cosmic rays, there is not sufficient time for a compound nucleus to
form and a dozen or more nucleons may escape and many combinations of
neutrons and protons are observed. Elementary particles such as mesons are
often produced.
4. *Fusion*—a heavier nucleus is built up from one or more lighter nuclei,
with a vast amount of energy released in the process. Gammas, positrons,
neutrinos, and/or neutrons may be emitted.

B. Threshold Energy and Q Energy

One of the most important factors that determine the type and probability of
a nuclear reaction is the kinetic energy of the incident particle. The Q energy
of a nuclear reaction

$$x + X \rightarrow (C) \rightarrow Y + y$$

is defined as

$$Q = E_Y + E_y = E_x \tag{1-1}$$

where E = kinetic energy
     x = incident particle or radiation
     X = target nucleus, assumed to be at rest
     Y = product nucleus
     y = outgoing particle or radiation
     C = compound nucleus.

    Since $(E_x + m_x c^2) + M_X c^2 = (E_Y + M_Y c^2) + (E_y + m_y c^2)$

where c = speed of light in vacuum and
   M, m = masses,
one has

$$Q = (M_X + m_x - M_Y - m_y)c^2 \tag{1-2}$$

depending only on the masses of the particles involved.

A necessary (but not always sufficient) condition for a nuclear reaction to take place is that the incident particle must have a minimum threshold energy, E_{th}, greater than or equal to the excitation energy of the compound nucleus. Thus, letting v_x and V_C be the velocities of the incident particle and compound nucleus, respectively,

$$E_{th} = \tfrac{1}{2}m_x v_x^2 = \tfrac{1}{2}M_C V_C^2 - Q$$

$$= -Q\left(1 + \frac{m_x}{M_X}\right) \tag{1-3}$$

since $m_x v_x = M_C V_C$ (conservation of momentum) and $M_C = M_X + m_x$ (definition of M_C).

C. Semiempirical Mass Formula and Binding Energy

Due to the complexity of an atomic nucleus, there are many models, each of which is suitable in certain particular cases. For the fission process that nuclear engineers are most interested in, the so-called liquid drop model appears to be most appropriate. According to this model, the nucleus has properties very much like that of a liquid droplet and one has the semiempirical mass formula for the mass or binding energy of a medium or heavy nucleus:

$$\text{B.E.} = a_v A - a_c \frac{Z(Z-1)}{A^{1/3}} - a_s A^{2/3} - a_a \frac{(A-2Z)^2}{A} + \frac{\delta}{2A} \tag{1-4}$$

where A = atomic mass of nucleus
Z = atomic number of nucleus
a's, δ = constants, determined empirically
B.E. = binding energy of a nucleus, defined by the relation:

$$M = ZM_p + (A-Z)M_n - \frac{\text{B.E.}}{c^2} \tag{1-5}$$

where M = mass of the nucleus
M_p = mass of a proton
M_n = mass of a neutron
c = speed of light in vacuum.

Each of the five terms on the right hand side of equation (1-4) has physical meaning:

1st term: "volume" energy proportional to A.

2nd term: "Coulomb" energy proportional to the number of "proton bonds", $Z(Z-1)/2$, and inversely proportional to R or $A^{1/3}$.

3rd term: "surface tension" energy proportional to R^2 or $A^{2/3}$.

4th term: "asymmetry" energy that would vanish if the neutron excess, $(A - 2Z)$, is zero.

5th term: "odd-even" energy that depends on whether Z or N are odd or even.

If the empirically determined values of the constants are substituted into equations (1-4) and (1-5), one obtains

$$\text{B.E.(Mev)} = 14.0A - 0.584\frac{Z(Z-1)}{A^{1/3}} - 13.1A^{2/3}$$

$$- 19.4\frac{(A-2Z)^2}{A} + \frac{\delta}{2A} \tag{1-6}$$

$$M\text{(amu)} = 0.99395A - 0.00084Z + 0.0141A^{2/3} + 0.021\frac{(A-2Z)^2}{A}$$

$$+ \frac{0.00063Z(Z-1)}{A^{1/3}} - \frac{\delta'}{A} \tag{1-7}$$

where

Z	N	δ(Mev)	δ'(amu)	A
even	even	270	0.145	even
even	odd	0	0	odd
odd	even	0	0	odd
odd	odd	−270	−0.145	even

and the *atomic* mass of 0^{16} is exactly 16 atomic mass units (amu). 1 amu = 931.145 Mev.

D. Radioactive Decay

Radioactive nuclei exist in nature and can also be produced by nuclear reactions such as nuclear transmutation or fission. In the case of simple radioactive decay, the number of radioactive nuclei at time t follows the familiar exponential decay rule:

$$N(t) = N_0\,e^{-\lambda t}$$

where the decay constant λ is the inverse of the average life of a nucleus, since

$$\text{average life} = \frac{\int_{N_0}^{0} t \, dN}{N_0} = \frac{\int_{\infty}^{0} t(-\lambda N_0 \, e^{-\lambda t}) \, dt}{N_0} = \frac{1}{\lambda}.$$

The decay constant is also related to the half-life T of a nucleus by

$$T = \frac{\ln 2}{\lambda} = \frac{0.6931}{\lambda}$$

In the case of sucessive decays, *i.e.*, if the parent nuclei N_1 decay into N_2 which in turn decay into N_3 and so on, the numbers of the various radioactive nuclei at time t can be obtained successively from a set of differential equations. For the particular case in which N_3 is stable, one writes:

$$\begin{cases} \dfrac{dN_1}{dt} = -\lambda_1 N_1 \\[2mm] \dfrac{dN_2}{dt} = \lambda_1 N_1 - \lambda_2 N_2 \\[2mm] \dfrac{dN_3}{dt} = \lambda_2 N_2 \end{cases}$$

from which one obtains:

$$N_1(t) = N_1^{\,0} \, e^{-\lambda_1 t} \tag{1-8}$$

$$N_2(t) = \frac{\lambda_1}{\lambda_2 - \lambda_1} N_1^{\,0} (e^{-\lambda_1 t} - e^{-\lambda_2 t}) + N_2^{\,0} \, e^{-\lambda_2 t} \tag{1-9}$$

$$N_3(t) = N_3^{\,0} + N_2^{\,0}(1 - e^{-\lambda_2 t})$$
$$+ N_1^{\,0}\left(1 + \frac{\lambda_1}{\lambda_2 - \lambda_1} e^{-\lambda_2 t} - \frac{\lambda_2}{\lambda_2 - \lambda_1} e^{-\lambda_1 t}\right) \tag{1-10}$$

where the superscript 0 indicates the value of N evaluated at $t = 0$. Successive radioactive decays are common in a reactor, as in the case of xenon-135 poisoning to be studied later. Figure 1.1 is a plot of N_1, N_2, and N_3 such that it is equivalent to the case in which a reactor is pulsed by a burst of neutrons

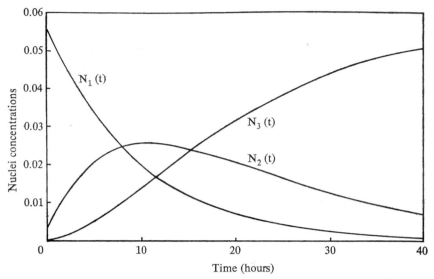

Figure 1.1   A plot of $N_1(t)$, $N_2(t)$, and $N_3(t)$ for $N_1{}^0 = 0.056$, $N_2{}^0 = 0.003$, $N_3{}^0 = 0$, $T_1 = 6.7$ hours, $T_2 = 9.2$ hours, and $T_3 = \infty$.

and N_2 is the xenon-135 concentration. The maximum occurs at about 11 hours after the burst.

SOLVED PROBLEMS

1-1   Give at least one example for each of the nuclear reactions discussed.
Answer:

Elastic scattering: $Au + \alpha \rightarrow Au + \alpha$ (Rutherford scattering)

Inelastic scattering: $Al^{27} + H^1 \rightarrow (Al^{27}) + H^1$

$$\phantom{Inelastic scattering: Al^{27} + H^1} \searrow Al^{27} + \gamma$$

Radioactive decay: $U^{236} \xrightarrow{\text{fission}} Y^{95} + I^{138} + 3n^1 + \text{energy}$

$$6.3\ s$$

$$Xe^{138} \xrightarrow[14\ m]{} Cs^{138} \xrightarrow[32.2\ m]{} Ba^{138}\ (\text{stable})$$

(*Note:* If a delayed neutron is given off, Ba^{137} might be the stable end product).

Radiative capture: $In^{115} + n^1 \rightarrow (In^{116}) \rightarrow In^{116} + \gamma$

$Al^{27}(p, \gamma)Si^{28}$

Fission: $n^1 + U^{233} \rightarrow (U^{236}) \rightarrow Ba^{141} + Kr^{92} + 3n^1, \quad Q = 200 \text{ Mev}$

$n^1 + U^{235} \rightarrow (U^{236}) \rightarrow Sr^{94} + Xe^{140} + 2n^1$

$n^1 + U^{235} \rightarrow (U^{236}) \rightarrow Ba^{139} + Kr^{94} + 3n^1$

Transmutation: $Be^9 + \gamma \rightarrow (Be^9) \rightarrow Be^8 + n^1, \quad Q = -1.67 \text{ Mev}$

$N^{14} + n^1 \rightarrow (N^{15}) \rightarrow C^{14} + H^1, \quad Q = 0.55 \text{ Mev}$

$Be^9(d, a)Li^7, \quad Q = 7.151 \text{ Mev}$

$U^{238} + O^{16} \rightarrow Fm^{250} + 4n^1$

Spallation: $p + p \rightarrow p + p + p + \bar{p}$ (6 Bev proton)

$p + \gamma \rightarrow \pi^0 + p$ \qquad (330 Mev γ)

Fusion: $H^2 + He^3 \rightarrow He^4 + H^1, \quad Q = 18.3 \text{ Mev}$

$H^2 + H^3 \rightarrow He^4 + n^1, \quad Q = 17.6 \text{ Mev}$

1-2 In the neutron producing $C^{12}(d, n)N^{13}$ reaction, what is the Q energy and what is the threshold energy?

Answer: The initial masses are:

$$C^{12} \text{ nucleus} = C^{12} \text{ atom} - 6e = 12.0038065 \text{ amu} - 6e$$
$$d = H^2 \text{ atom} - \quad e = \underline{2.0147361 \text{ amu} - \quad e}$$
$$14.0185426 \text{ amu} - 7e$$

The final masses are:

$$N^{13} \text{ nucleus} = N^{13} \text{ atom} - 7e = 13.0098617 \text{ amu} - 7e$$
$$n = \underline{1.0089830 \text{ amu}}$$
$$14.0188447 \text{ amu} - 7e$$

$Q = $ (initial mass—final mass) $931.145 = -0.282 \text{ Mev}$

The threshold bombarding energy for deuterons is

$$E_{th} = +0.282 \left(1 + \frac{2}{12}\right) = 0.328 \text{ Mev}$$

Thus, the neutron produced has energy from zero upward, depending on the energy of the deuteron in excess of 0.328 Mev. If C^{12} ions were used to bombard a deuterium target, the threshold energy would have been

$$E_{th} = +0.282 \left(1 + \frac{12}{2}\right) = 1.97 \text{ Mev}$$

Note: a) The atomic mass unit used in this book is based on a scale that an O^{16} atom has exactly 16 amu (1 amu $= 931.145$ Mev). A list of the masses of the elements and their isotopes is given by Mattauch and Everling, and Duckworth in "Progress in Nuclear Physics", O. R. Rrisch, Ed., Vol. 6, P. 138 & P. 233, (1957).

b) The popular "Chart of the Nuclides" by Knolls Atomic Power Laboratory uses a scale that a C^{12} atom has exactly 12 amu. (1 amu = 931.441 Mev).

c) 1 amu (C-12) = 1.00031792 amu (O-16).

1-3   Calculate the Q energy and threshold energy of the $d(d, n)He^3$ reaction which is used to produce high energy neutrons.
Answer: The initial masses are, including 2e, 4.0294722 amu.
The final masses are, including 2e, 4.0259810 amu.
Thus, $Q = +3.25$ Mev.
Since Q is positive, the nuclear reaction is not limited by threshold energy. The minimum neutron energy is obtained when the deuterons have zero initial kinetic energies and the neutron and He^3 are emitted in opposite direction such that the total momentum is zero. In other words, the neutron will have at least 2.44 Mev.

1-4   In a nonrelativistic $X(x, y)Y$ reaction, if y and Y are emitted with angles θ and ϕ, respectively, relative to the direction of the incident particle x, obtain an expression for Q that does not depend on E_Y, since the latter is usually very small and difficult to measure experimentally.
Answer: Assume $V_X = 0$.

Conservation of momentum: $\begin{cases} m_x v_x = M_Y V_Y \cos\phi + m_y v_y \cos\theta \\ 0 = -M_Y V_Y \sin\phi + m_y v_y \sin\theta \end{cases}$

Eliminating ϕ by squaring and adding the two equations, one has, by changing the velocities to kinetic energies and rearranging terms,

$$E = \frac{m_x}{M_Y} E_x + \frac{m_y}{M_Y} E_y - \frac{2}{M_Y} (m_x m_y E_x E_y)^{\frac{1}{2}} \cos\theta$$

Since $Q = (E_Y + E_y) - E_x$, one has

$$Q = E_y \left(1 + \frac{m_y}{M_Y}\right) - E_x \left(1 - \frac{m_x}{M_Y}\right) - \frac{2}{M_Y} (E_x E_y m_x m_y)^{\frac{1}{2}} \cos\theta \qquad (1\text{-}11)$$

as the general equation for the Q value of a nuclear reaction.

1-5   In the $B^{11}(d, \alpha)Be^9$ reaction, the α was observed to have 6.37 Mev at $\theta = 90°$. What must be the energy of the deuteron?
Answer:

$$m_x = 2.014736 \text{ amu} - e \qquad m_y = 4.0038727 \text{ amu} - 2e$$
$$M_X = 11.012795 \text{ amu} - 5e \qquad M_Y = 9.015041 \text{ amu} - 4e$$

The sum of the initial masses and the sum of the final masses of the particles have a difference of 0.0086173 amu. Thus, $Q = +8.02$ Mev.
From equation (1-11),

$$8.02 = 6.37(1 + 0.444) - E_x(1 - 0.222) - 0$$

or,

$$E_x = 1.505 \text{ Mev}$$

The same technique can be used to measure neutron energy.

1-6 Calculate the binding energy and mass of U-238 atom from the semiempirical formula.
Answer:

$$A = 238 \text{ (even)}, \ Z = 92 \text{ (even)}, \ \delta' = +270, \ \delta = +0.145.$$

From equations (1-6) and (1-7), one obtains
B.E. = 1819 Mev/nucleus = 7.65 Mev/nucleon
M = 238.1321 amu (O-16 scale)
in comparison with the experimentally determined and accepted values of
B.E. = 1803 Mev/nucleus = 7.58 Mev/nucleon and M = 238.1243 amu.

1-7 Prove that for beta decay to occur, the mass of the parent atom must be greater than that of the daughter atom.
Answer: Consider $_Z X^A \rightarrow {}_{Z+1} Y^A + {}_{-1}e^0$ and $Q = (M_X - M_Y - m_e)c^2$ where M_X and M_Y are nuclear masses. For spontaneous decay, one must have $Q > 0$. Thus, $M_X > M_Y + m_e$. However,

$$M(Z) = M_X + m_e Z$$

$$M(Z + 1) = M_Y + m_e(Z + 1)$$

where $M(Z)$ and $M(Z + 1)$ are the atomic masses of the parent and daughter atoms, respectively. Hence,

$$M(Z) > M(Z + 1)$$

is a necessary condition. An example of this is that after U-235 captures a slow neutron and changes to U-236, the U-236 cannot beta decay into Np-236. On the other hand, U-238 may capture a neutron and beta decay eventually to Pu-239.

SUPPLEMENTARY PROBLEMS

1-a Is it theoretically possible to bombard stationary Be^9 with 2 Mev protons and get neutrons through the Be$^9(p, n)$B^9 reaction?

1-b Calculate the binding energy per nucleon of U-236, Te-135, and Zr-99. If the latter two are fission fragments of U-236, what is the calculated Q energy of the fission reaction?

1-c Derive conditions for the occurrences of positron emission $(_Z X^A \rightarrow {}_{Z-1} Y^A + {}_{+1}e^0)$ and orbital electron capture $(_Z X^A + {}_{-1}e^0 \rightarrow {}_{Z-1} Y^A)$ analogous to that in problem (1-7).

Chapter 2

NEUTRON SOURCES AND DETECTION

NUCLEAR fissions are induced by neutrons. In order to study the interactions of neutrons with matter, it is desirable to study the generation and detection of neutrons. A few useful terminologies are first introduced.

A. Definitions

1. Maxwellian Distribution

When a collection of particles, such as neutrons, is in thermal equilibirum with a substance at temperature $T°K$, the speed and energy distributions are given by the distribution equations of Maxwell:

$$\frac{n(v)}{n_0} = 4\pi \left(\frac{m}{2\pi kT}\right)^{3/2} v^2 \, e^{-mv^2/(2kT)}$$

$$\frac{n(E)}{n_0} = \frac{2\pi}{(\pi kT)^{3/2}} \, e^{-E/(kT)} E^{1/2}$$

where n_0 = total number of neutrons

$n(v)$ = number of neutrons of speed v per unit speed interval

$n(E)$ = number of neutron of energy E per unit energy interval

m = mass of neutron = 1.675×10^{-24} g.

k = Boltzmann constant

= 1.3804×10^{-16} erg/degree

= 8.61×10^{-5} ev/degree

2. Thermal Neutron

Neutrons in thermal equilibrium with a substance are called thermal neutrons. Unless otherwise specified, the temperature is taken as the room temperature of $293°K$. The average kinetic energy (not speed) of the neutrons is the same as that of the atoms or molecules of the medium. For thermal neutrons in a Maxwellian distribution,

the most probable speed $\qquad v_p = \sqrt{\frac{2kT}{m}} = 2200$ m/sec $\cong 1.3$ miles/sec the average speed $\qquad \bar{v} = \sqrt{\dfrac{8kT}{\pi m}} = 1.128v_p$

the root-mean-square speed $\quad v_{\text{rms}} = \sqrt{\dfrac{3kT}{m}} = 1.224v_p$

the average kinetic energy $\quad \bar{E} = \frac{1}{2}mv_{\text{rms}}^2 = \frac{3}{2}kT$
the most probable energy $\qquad E_p = \frac{1}{2}kT \neq \frac{1}{2}mv_p^2$

the energy corresponding to the most probable speed is
$$E = \tfrac{1}{2}mv_p^2 = kT = 0.0253 \text{ ev.}$$

Nuclear data for thermal neutrons are given for 2200 m/s (0.0253 ev) neutrons, unless otherwise specified.

3. *Cross-section*

The cross-section of a nuclear reaction is a measure of the probability of occurrence of that reaction. For any collision reaction between nuclear particles or radiation, the microscopic cross-section σ may be looked upon as an area such that the number of reactions taking place is equal to the product of the number of incident particles that would pass through this area at normal incident and the number of target nuclei. The unit for σ is *barn*, 10^{-24} cm^2. The macroscopic cross-section Σ is defined by

$$\Sigma = N\sigma$$

where N is the number of nuclei per c.c. The unit for Σ is cm^{-1}.

There are many types of cross-sections. Examples are fission cross-section σ_f, radiative capture cross-section σ_r, absorption cross-section σ_a, and scattering cross-section σ_s. In general,

$$\sigma_a = \sigma_f + \sigma_r$$

while the total cross-section σ_t is given by $\sigma_a + \sigma_s$. Neutrons of intermediate energies also may experience resonance scattering and absorption by many nuclei. It is the phenomenon of relatively high cross-sections over isolated narrow ranges of energy. For U^{235}, the fission cross-section has resonances between 1 ev and 1 kev.

4. *Flux*

The ratio of the reaction rate per c.c., R, and the macroscopic cross-section of that reaction, Σ, is called flux, ϕ.

$$\phi = \frac{R}{\Sigma} = \frac{nv\Sigma}{\Sigma} = nv$$

B. Neutron Sources

In many experiments, such as that in the study of nuclear materials and the measurements of nuclear cross-sections, a neutron source is required. A list of the methods of neutron production is given below:

1. *(α, n) reaction with light elements:* The alphas may be from Ra^{226}, Po^{210}, or Rn^{222}. The light elements are Be, Li, F, or B. Neutron energies from about 1 Mev to about 13 Mev may be otained. The three alpha sources are compared below:

	$T_{\frac{1}{2}}$	Advantages	Disadvantages
$_{84}Po^{210}$	140 d	low gamma flux	short $T_{\frac{1}{2}}$; low neutron emission rate.
$_{86}Rn^{222}$	3.8 d	Gas; compactness	short $T_{\frac{1}{2}}$
$_{88}Ra^{226}$	1620 yr	Long $T_{\frac{1}{2}}$; large neutron emission rate	large gamma flux

The most useful reaction is that from Be-Ra source:

$$Be^9 + He^4 \rightarrow C^{12} + n^1 + 5.71 \text{ Mev}$$

From a mixture of one curie (3.7×10^{10} disintegrations/sec) of radium (exactly 1 gram in this case) with Be powder, about 1.7×10^7 neutrons per second is obtained.

2. *(γ, n) reaction with Be and H^2:* The reactions are:

$$Be^9(\gamma, n)Be^8, \qquad E_{th} = 1.67 \text{ Mev}$$

$$d(\gamma, n)p, \qquad E_{th} = 2.23 \text{ Mev}$$

The neutrons thus produced are often called photoneutrons. If the gammas are from a reactor, the neutron source can be made very intense. If the gammas from radioactive sources such as Na^{24}, Mn^{56}, Ga^{72}, In^{116}, Sb^{124}, and La^{140} are used, the photoneutrons will be essentially monoenergetic. It may be noted that the gamma sources listed all have $T_{\frac{1}{2}}$ less than 60 days and the photoneutrons have average energy of less than 1 Mev.

Source	$T_{\frac{1}{2}}$	Average neutron energy, kev	Neutron per second per curie, with 1 gm of target at 1 cm
$Na^{24} + D_2O$	14.8 hr	220($\pm$20)	2.7×10^5
$Na^{24} + Be$	14.8 hr	830($\pm$40)	1.3×10^5
$Sb^{124} + Be$	60 d	30	1.9×10^5

3. *(d, n) and (p, n) reactions with light elements:* With deuterons and protons produced in particle accelerators as incident particles, the following reactions

serve as neutron sources:

$$D(D, n)He^3 + 3.28 \text{ Mev}$$
$$T(D, n)He^4 + 17.6 \text{ Mev}$$
$$T(p, n)He^3 - 0.764 \text{ Mev}, \qquad E_{th} = 1.02 \text{ Mev}$$
$$Li^7(p, n)Be^7 - 1.65 \text{ Mev}, \qquad E_{th} = 1.89 \text{ Mev}$$
$$Be^9(d, n)B^{10} + 4.35 \text{ Mev}$$
$$C^{12}(d, n)N^{13} - 0.26 \text{ Mev}, \qquad E_{th} = 0.32 \text{ Mev}$$

The first four reactions are plotted in Figure 2.1. The second reaction above indicates that neutrons with energies up to 20 Mev may be obtained when

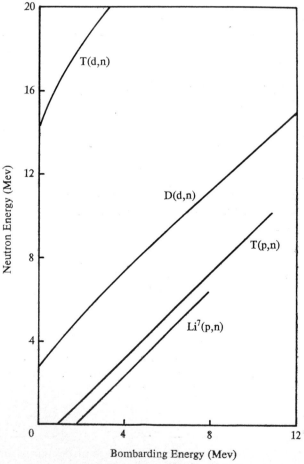

Figure 2.1 Energies of neutrons emitted in the forward direction.

tritium is bombarded with low energy deuterons. Depending on the angles of emission, the neutrons are quite monoenergetic.

4. *Stripping of deuterons:* When a beam of very high energy deuterons strikes a target, the proton and neutron of the deuteron may be separated since they are only loosely bound by about 2.23 Mev. As a result, the neutron may proceed forward with about half of the deuteron energy and the proton may be absorbed by the target nucleus. Neutrons produced in this manner may have energies of a few hundred Mev.

5. *High energy (p, n) reaction:* Neutrons of energy up to a few Bev may be obtained by letting a high energy proton hit a neutron of a target nucleus head-on. In such a head-on collision, all the energy and momentum of the incoming proton are transferred to the outgoing neutron.

6. *Fission neutrons from a nuclear reactor:* Neutron fluxes of the order of 10^{14} neutrons/cm^2-sec are very common in a reactor, in comparison with that of about 10^8 n/cm^2-sec from the accelerators with the $H^2(d, n)He^3$ reaction. Fission neutrons have energies up to about 18 Mev and have average energy of around 2 Mev.

7. *Thermal neutrons by slowing down:* Thermal neutrons may be obtained by slowing down fast neutrons through elastic and inelastic scatterings. If a 2 Mev neutron source is surrounded by water, the thermal neutron flux is at a maximum about 10 cm away from the source.

8. *Californium* 252: Through successive neutron capturing transmutations, Cf^{252} may be produced. The analogy of successive transmutation may be that of a typical spectator catching four home-run baseballs in the stands during his lifetime. Even with a high flux reactor, such as the high flux isotope production reactor (HFIR) at Oak Ridge, only a small amount of Cf^{252} can be obtained as the result of a yearround irradiation program. The isotope is a very intense neutron emitter and is being used in research. In passing, it may be noted that the neutron flux in the HFIR is so intense that a new core has to be replaced every three months or so.

C. Neutron Detection and Energy Determination

Since neutrons produce very little direct ionization in their passage through a material, they are detected by the indirect ionization which results from some nuclear reaction. The methods for neutron flux and energy measurements are:

1. BF$_3$ *Counter*

For slow neutrons (of kinetic energy less than about 1 ev), it is convenient to utilize the reaction

$$B^{10} + n \rightarrow Li^7 + \alpha + 2.79 \text{ Mev} \quad (6.3\% \text{ for thermal neutrons})$$

or,

$$B^{10} + n \rightarrow (Li^7) + \alpha + 2.31 \text{ Mev}$$
$$\phantom{B^{10} + n \rightarrow } \hookrightarrow Li^7 + \gamma + 0.48 \text{ Mev}$$

Natural boron contains 19.78% of B^{10}. The thermal neutron cross-section for the reaction above is 4020 barns. It is because of the high cross-section that make B^{10} an useful neutron detector. (However, the cross-section varies inversely with the velocity of the neutrons up to energies of about 1 kev). Thus, if the inner wall of a proportional counter tube is lined with boron-rich material or, as is more often done, if a proportional counter tube filled with boron trifluoride (BF_3) gas or a boron loaded scintillation counter is used, the ionizing alphas will indicate the presence of the neutrons.

2. *Long Counter*

For fast and slow neutron flux determination, the so-called long counter is effective. It is nothing more than a BF_3 counter surrounded by about 3" of paraffin. The paraffin, being rich in hydrogen, is effective in slowing down the fast neutrons.

Stray neutrons from various directions other than the incoming neutron beam direction can be picked up by another 3" of paraffin outside of $\frac{1}{2}$" of B_2O_3. For fast neutron flux determination only, the counter may be shielded by a thin sheet of cadmium which has very high thermal neutron absorption cross-section.

The long counter (which is generally no more than 18" long) has the characteristics of being almost energy independent over a long range of up to about 5 Mev.

The flux determination efficiencies of both the BF_3 counter and the long counters can be calibrated using a neutron source of known strength.

3. *Time-of-Flight Method and Neutron Chopper*

For neutron velocity (and hence energy) determination, the time-of-flight method is popular and accurate. A continuous beam of neutrons is first passed through a neutron "chopper" or velocity selector. It consists of a rotating disc made of stacks of cadmium and aluminum plates. The neutron absorbing cadmium acts as a shutter. When bursts of neutrons emerge from the chopper, they are allowed to travel a distance L before they reach a detector. The time t required for the flight is obtained by synchronizing the source and detector signals. Obviously,

$$E = \tfrac{1}{2}mv^2 = \tfrac{1}{2}m(L/t)^2$$

Neutron energies ranging from fractions of an ev to about 150 Mev can be determined accurately by this method, using nanosecond resolving time electronic circuits. In many cases, gamma rays are emitted in conjunction with the study of nuclear reactions. These gammas may be used as the base for the timing circuit of the synchronized detectors.

4. Neutron Crystal Spectrometer

Based on the properties of neutron diffraction by a crystal, this method is very accurate in the measurement of neutron energies, but is only good for slow neutrons. According to Bragg's condition:

$$n\lambda = 2d \sin \theta$$

where d = spacing between different planes of the atoms in a crystal
n = order of diffraction
θ = half the angle between the incident neutron beam and the diffracted neutron beam
λ = de Broglie wavelength of neutron
$= h/(2mE)^{\frac{1}{2}}$

where h = Planck constant = 6.625×10^{-27} erg-sec.
m = neutron rest mass = 1.6748×10^{-24} g
E = energy of neutron.

A neutron crystal spectrometer can be used for:
a) Determining or selecting E, knowing d, n, and θ.
b) Studying crystal structure, knowing E, n, and θ.
c) Studying the energy distribution of neutrons by considering the strength of the diffracted neutron beam at various angles.

It may be noted that a thermal neutron of 0.0253 ev has $\lambda = 1.8$ Å which is of the same order of magnitude of d for most crystals.

5. Activation Analysis

Many nuclei, such as In^{115}, Au^{197}, Mn^{55}, Co^{59}, Cu^{63}, Cu^{65}, Lu^{175}, and Lu^{176}, become radioactive by capturing neutrons. By studying their decays, it is possible to get information about the original neutron fluxes. Start from

$$\frac{dN_a(v)}{dt} = n(v)v\sigma_a(v)N$$

where $N_a(v)dv$ = the number of new nuclei per c.c. produced by absorbing neutrons of velocities between v and $v + dv$
$n(v)dv$ = number of neutrons per c.c. having velocities between v and $v + dv$

c

$\sigma_a(v)$ = neutron absorption cross-section at v

N = the number of old nuclei per c.c. in the target sample which can undergo the given reaction.

In most cases, the absorption cross-section $\sigma_a(v)$ follows the $1/v$ law. *i.e.*,

$$\sigma_a(v)v = \sigma_{a0}v_0$$

where σ_{a0} is the absorption cross-section at some particular velocity v_0, such as 2200 m/s. Then,

$$\frac{dN_a(v)}{dt} = n(v)\sigma_{a0}v_0 N$$

or, integrating over the entire velocity range,

$$\frac{dN_a}{dt} = n\sigma_{a0}v_0 N$$

where N_a = number of new nuclei per c.c. formed by the absorption of neutrons of any velocity

n = total number of neutrons per c.c.

If N_a decays with a decay constant λ, *i.e.*,

$$\frac{dN_a}{dt} = nv_0\sigma_{a0}N - \lambda N_a$$

then the *activity* $A(\tau)$ measured at a time τ after the irradiation is

$$A(\tau) = \lambda N_a\,e^{-\lambda\tau} = nv_0\sigma_{a0}N(1 - e^{-\lambda t})\,e^{-\lambda\tau} \quad \text{disintegrations/sec} \qquad (2\text{-}1)$$

where t is the length of time for the neutron irradiation of the sample. The equation above is true for any neutron energy distribution, including that of Maxwellian distribution. It may be used to determine the neutron density n when the activation cross-section σ_{a0} is known, or vice versa.

6. Threshold Detector

We have seen in Chapter 1 that certain nuclear transmutations require a minimum threshold energy. In the case that the recoil nucleus is radioactive, one can, by measuring the activity of the radioactive nucleus, estimate the flux of neutron with energy above the threshold. This method is usually good for fast neutrons.

Isotope	Approximate threshold energy (Mev)	Effective cross-section (barns)
P^{31}	2	0.07
Al^{27}	4	0.08
Cu^{63}	12	0.9
Ni^{58}	13	1.1
C^{12}	22	0.2

7. Fission Chamber

By lining the interior of an ionization chamber with fissionable materials such as U^{235} (for slow neutrons), U^{238}, and Th^{232} (for fast neutrons), it is possible to detect neutrons from the large pulses caused by the energetic fission fragments. Fast and slow neutron fluxes can be obtained separately with the help of a thin cadmium thermal neutron shield.

8. Proton Recoil Counter

Fast neutrons may be counted by an ionization chamber filled with hydrogenous gas. Incoming fast neutrons transfer energy by elastic collisions to the protons and the latter produce ionizations that give rise to pulses. Slow neutrons cannot be detected efficiently in this manner, since the recoil energy of the protons is too small to cause appreciable ionization.

9. Chemical Detectors

The amount of darkening of a photographic emulsion is dependent upon the flux and energy of the accumulative radiation and photographic print-out emulsions that contain excess silver ions and that do not require wet processing are used extensively in the film badges.

The structure of many plastics is altered by radiation, due to the cross linkage of polymers. Changes in transparency, color, electrical conductivity, density, and mechanical properties can be calibrated against known radiation sources and used as detectors. The Fricke method, for example, uses an aqueous solution of 0.001 M $FeSO_4$, 0.001 M NaCl, and 0.4 M H_2SO_4 as the basis of a chemical technique in measuring the color changes due to radiation.

10. Others

There are other types of neutron detectors, such as neutron sensitive semiconductor detectors.

11. Compensated Ion Chamber

To compensate for or to substract away the effect of a gamma flux that is usually present with a neutron flux, a compensated ion chamber may be used. It consists of an ionization chamber with two regions. One region contains boron-10 so that it is sensitive to both neutrons and gammas. The other region contains no boron-10 and is only gamma sensitive. The volumes and voltages of the two regions are such that the outputs from the two are identical in a gamma field alone. In a neutron and gamma field, the difference of the two outputs, substracted electronically, gives an indication of the neutron flux alone and can be calibrated. Compensated ionization chambers are used for monitoring the neutron flux in a reactor.

SOLVED PROBLEMS

2-1 Prove that the average speed $\bar{v}$ of a Maxwellian distribution is $\sqrt{8kT/\pi m}$.
Answer:

$$\bar{v} = \frac{\int_0^\infty n(v)\, v\, dv}{\int_0^\infty n(v)\, dv} = \frac{\int_0^\infty v^3\, e^{-mv^2/(2kT)}\, dv}{\int_0^\infty v^2\, e^{-mv^2/(2kT)}\, dv} = \sqrt{\frac{8kT}{\pi m}}$$

2-2 Calculate the macroscopic absorption cross-section of natural uranium (density = 18.7 g/cc) containing 0.7% U-235 and 99.3% U-238 for 2200 m/s neutrons.
Answer:

$$\Sigma_a = N\sigma_a = \rho\, \frac{A_0 \sigma_a}{A} = 18.7\, A_0 \left(\frac{0.007\, \sigma_a^{235}}{235} + \frac{0.993\, \sigma_a^{238}}{238} \right)$$

where A_0 = Avogadro's number = 0.602472×10^{24} (gram-mole)$^{-1}$.

From the Chart of the Nuclides (1964), one obtains:

$$\sigma_a^{235} = \sigma_r + \sigma_f = 100 + 580 = 680 \text{ barns}$$

$$\sigma_a^{238} = 2.7 \text{ barns}$$

Thus,
$$\Sigma = 18.7 \times 0.602 \left(\frac{0.007 \times 680}{235} + \frac{0.993 \times 2.7}{238} \right)$$

$$= 11.24\, (0.0202 + 0.0116) = 0.358 \text{ cm}^{-1}.$$

2-3 In the time-of-flight experiment, if $L = 3$ meters, calculate the times of flight for a) 0.2 ev. b) 20 Mev neutrons.
Answer:

$E = \tfrac{1}{2}m(L/t)^2$, $m = 1.6748 \times 10^{-24}$ g., 1 ev = 1.60207×10^{-12} erg.

a) $0.2 \times 1.6 \times 10^{-12} = \frac{1}{2}(1.6748 \times 10^{-24})(300/t)^2$

$$t = 485 \text{ microseconds}$$

b) $t = 48.5$ nanoseconds.

2-4 A thin 1.5 gram natural copper foil, which is an $1/v$ absorber (*i.e.*, $\sigma_a \propto 1/v$), is exposed to a Maxwellian neutron flux of 10^{14} neutrons/cm²-sec. What is the production rate of Cu^{64}?

Answer: Natural copper contains 69% of Cu^{63}. The number of Cu^{63} atoms per foil = $1.5[(0.602 \times 10^{24})/63]0.69 = 0.00987 \times 10^{24}$.

The production rate $R = 10^{14} \times 0.00988 \times 10^{24} \times \bar{\sigma}$
where $\bar{\sigma}$ is the effective absorption cross-section of Cu^{63} over the entire velocity range. The flux of 10^{14} is given for the entire velocity range, not only at 2200 m/s). Now,

$$\bar{\sigma} = \frac{\int_0^\infty n(v)v\sigma(v)\,dv}{\int_0^\infty n(v)v\,dv} = \frac{\int_0^\infty n(v)v_p\sigma_p\,dv}{\int_0^\infty n(v)v\,dv} = \frac{v_p\sigma_p\int_0^\infty n(v)\,dv}{\int_0^\infty n(v)v\,dv},$$

or

$$\bar{\sigma} = \frac{v_p\sigma_p}{\bar{v}} = \sigma(\bar{v}) = \frac{\sigma_p}{1.128} \tag{2-2}$$

where $v_p = 2200$ m/s and $\sigma_p = \sigma(v_p) = 4.5$ barns. Thus,

$$R = 10^{14} \times 0.00987 \times 10^{24} \times \frac{4.5 \times 10^{-24}}{1.128} = 3.94 \times 10^{12} \text{ atoms/foil-sec.}$$

2-5   If the foil in problem (2-4) above is exposed to the flux for 2 hours and then removed, what is the activity of the foil 5 hours after removal?

Answer: Natural copper contains 69% of Cu^{63} and 31% of Cu^{65}. The absorption cross-sections are 4.5 barns and 2.3 barns, respectively. The half-lives of Cu^{64} and Cu^{66} are 12.9 hr and 5.1 m, respectively. Thus, 5 hours after removal, the Cu^{66} would be negligible in concentration in comparison to the Cu^{64}. Referring to equation (2-1) and the result obtained above,

$t = 2$ hr, $\tau = 5$ hr, $\lambda = 0.6931/12.9 = 0.0537$ hr^{-1}.

$$A(\tau) = 3.94 \times 10^{12} (1 - e^{-0 \cdot 0537 \times 2}) e^{-0 \cdot 0537 \times 5}$$

$$= 3.94 \times 10^{12}(1 - 0.8981)(0.7645)$$

$$= 0.306 \times 10^{12} \text{ disintegrations per foil per second.}$$

SUPPLEMENTARY PROBLEMS

2-a   Suggest a method for producing monoenergetic neutrons in each of the following energy ranges: 10^{-3} ev, 10^2 ev, 10^7 ev, and 10^9 ev.

2-b Calculate Σ_a for H_2O and D_2O.

2-c In measuring the neutron energy by crystal spectrometer, the uncertainty $\Delta\theta$ in the Bragg angle produces uncertainty in the energy. Derive an expression for the fractional uncertainty $\Delta E/E$ in terms of θ and $\Delta\theta$.

2-d In problem (2-5), what percentage of the Cu^{63} atoms in the foil were "burned-up"?

Chapter 3

FISSION

IN THIS chapter, the most basic features of the fission process are presented.

A. Fissionable Materials—General Statements

1. Fission can be produced in uranium and thorium not only by neutron bombardment but also by high-energy alphas, protons, deuterons, and gammas. Fission can also be produced in other nuclides under different conditions, although a chain reaction cannot always be assured.

2. U-235 is the only *natural* occurring element that can be fissioned efficiently by thermal neutrons. (U-233, for example, can also be fissioned by thermal neutrons, but its *natural* abundance is negligibly small.)

3. Naturally occurring U-238 and Th-232 undergo fission when bombarded with fast neutrons of energies greater than 0.7 and 1.7 Mev, respectively. At lower neutron energies, (n, γ) reactions followed by β-decays result in Pu-239 and U-233, both fissionable when bombarded with slow or fast neutrons.

$$_{92}U^{238} + n \longrightarrow {}_{92}U^{239} \qquad\qquad {}_{90}Th^{232} + n \longrightarrow {}_{90}Th^{233}$$

$$_{92}U^{239} \xrightarrow{\;23.5m\;} {}_{93}Np^{239} + e \qquad {}_{90}Th^{233} \xrightarrow{\;23.3m\;} {}_{91}Pa^{233} + e$$

$$_{93}Np^{239} \xrightarrow{\;2.35d\;} {}_{94}Pu^{239} + e \qquad {}_{91}Pa^{233} \xrightarrow{\;27.4d\;} {}_{92}U^{233} + e$$

4. Natural occurring uranium contains:

	Weight per cent	Isotopic mass (0–16 scale)
U-238	99.283	238.1252
U-235	0.711	235.1175
U-234	0.0058	234.1140
Others	0.0002	

B. Fission Energy

The approximate distribution of fission energy released from the thermal neutron fission of an U^{235}, Pu^{239}, or U^{233} atom is given below:

	Mev
Kinetic energy of fission fragments	165
Prompt gammas from excited fission fragments	7
Kinetic energy of fission neutrons	5
Beta particles from fission product decays	7
Delayed gammas from fission product decays	6
Neutrinos from beta decays	10
	200

The 10 Mev carried away by the highly penetrating neutrinos does not contribute to the practical utilization of fission energy. It is convenient to note the following conversion relations:

1 metric ton = 1 tonne = 10^6 g.

1 Mev = 1.60×10^{-6} erg = 1.60×10^{-13} watt-sec.

1 BTU = 251.8 calories = 1054.8 watt-sec.

          = 2.93×10^{-4} kilowatt-hour = 3.93×10^{-4} horsepower-hour

Avogadro number = 0.602×10^{24} atoms/gram-mole

and, based on 190 Mev/fission, the following relations may be obtained:

190 Mev/fission

3.04×10^{-11} watt-sec/fission

3.3×10^{10} fission/watt-sec.

1.3 grams of U^{235} consumed per megawatt-day

1.0×10^7 kw-hr/lb of U^{235} fissioned.

C. Experimental Data

Figures 3.2 through 3.6 are experimental data on the fission neutrons and fission fragments. Figures 3.7 and 3.9 are fission cross-section of U^{235} and total cross-section of U^{238}, respectively, while Table 3.1 is a list of thermal neutron cross-sections. The prompt fission gamma energy spectrum is given in Figure 3.11. Figures 3.1 and 3.8 are for convenient references and comparisons. Figure 3.10 will be discussed later in this chapter.

D. Prompt Neutrons

Prompt neutrons are often referred to as fission neutrons, since over 99% of the neutrons produced by the fissions of U^{235}, Pu^{239}, and U^{233} are emitted

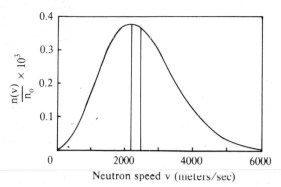

Figure 3.1 Maxwellian distribution of neutrons in medium at 293°K.

$$v_p = 2200, \bar{v} = 2482.$$

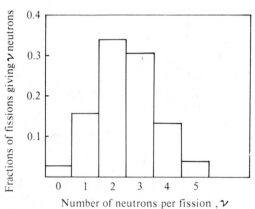

Figure 3.2 Fission neutron number distribution due to thermal fission of U^{235}.

$$\bar{\nu} = 2.44 \pm 0.02$$

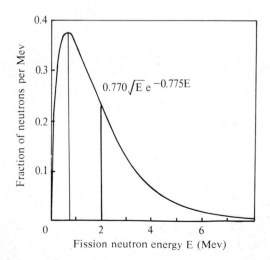

Figure 3.3 Fission neutron energy distribution due to thermal fission of U^{235}.

$$E_p = 0.72, \bar{E} = 2.$$

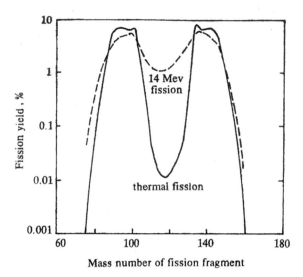

Figure 3.4 Fission yields of U^{235} as function of mass number.

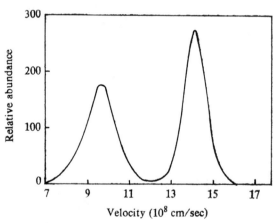

Figure 3.5 Distribution in velocity of the fission fragments from the thermal fission of U^{235}.

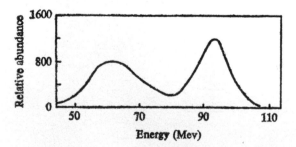

Figure 3.6 Energy distribution of the fission fragments from the thermal fission of U^{235}.

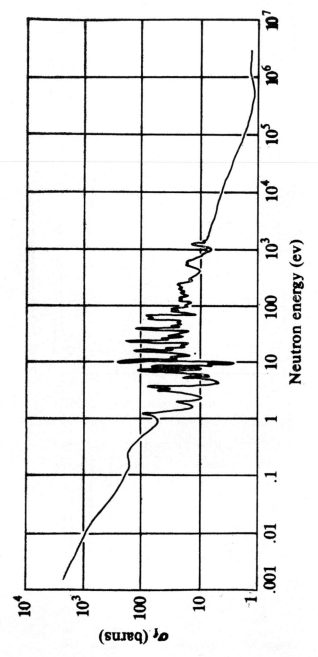

Figure 3.7 Fission cross-section of U^{235}.

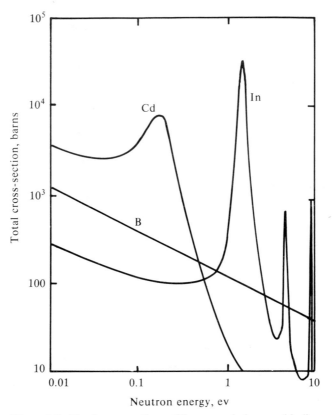

Figure 3.8   Total cross-sections of boron, cadmium, and indium.

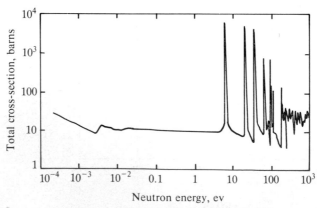

Figure 3.9   Total cross-section of uranium-238.

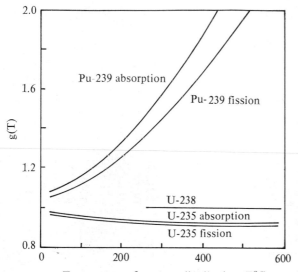

Figure 3.10 Non-1/v factor $g(T)$ as a function of temperature.

$$\bar{\sigma}(T) - g(T)\frac{\sigma(T_0)}{1.128}\left(\frac{T_0}{T}\right)^{1/2} ;$$

T and T_0 in °K.

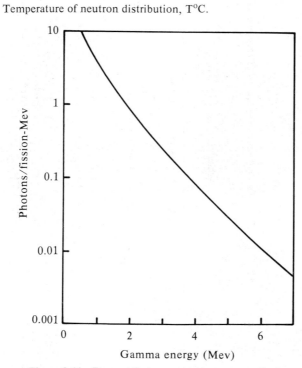

Figure 3.11 Prompt fission gamma energy spectrum.

TABLE 3.1 Thermal neutron cross-sections of certain elements, compounds, and isotopes.

	Atomic or molecular wt. (amu)	Density (g/c.c.)	Nuclei per c.c. $\times 10^{-24}$	σ_a (barns)	σ_s (barns)	Σ_a (cm^{-1})	Σ_s (cm^{-1})	σ_f (barns)
H	1.008	0.000089	0.000053	0.330	38	0.000017	0.002	
H_2O	18.016	1	0.0335*	0.66	103	0.022	3.45	
D_2O	20.030	1.10	0.0331*	0.001	13.6	0.000033	0.449	
Li	6.940	0.534	0.0463	71	1.4	3.29	0.065	
Li-6				945 (n, α)				
Li-7				0.033 (n, γ)				
Be	9.013	1.85	0.1236	0.010	7.0	0.00124	0.865	
BeO	25.02	3.025	0.0728*	0.010	6.8	0.00073	0.501	
B	10.82	2.45	0.1364	755	4	103	0.346	
B-10				3813 (n, α)				
B-11				<0.05 (n, γ)				
C	12.011	1.60	0.0803	0.004	4.8	0.00032	0.385	
graphite	12	1.70	0.0855	0.0037	4.8	0.00032	0.41	
N	14.008	0.0013	0.000053	1.88	10	0.000099	0.00050	
O	16.000	0.0014	0.000053	0.00020	4.2	0.000	0.00021	
F	19.00	0.0017	0.000053	0.01	3.9	0.0000005	0.00020	
Na	22.991	0.971	0.0254	0.525	4	0.013	0.102	

Al	26.98	2.699	0.0602	0.241	1.4	0.015	0.084
Fe	55.85	7.86	0.0848	2.62	11	0.222	0.933
Zr	91.22	6.4	0.0423	0.185	8	0.008	0.338
ZrH_2	93.2	5.61	0.036	0.84		0.030	
Cd	112.41	8.65	0.0464	2450	7	114	0.325
In	114.82	7.28	0.0382	191	2.2	7.30	0.084
I	126.91	4.93	0.0234	7.0	3.6	0.164	0.084
Xe	131.30	0.0059	0.000027	35	4.3	0.00095	0.00012
Xe-135				$2.7 \times 10^6\ (n, \gamma)$			
Au	197	19.32	0.0591	98.8	9.3	5.79	0.550
Pb	207.21	11.35	0.0330	0.170	11	0.006	0.363
Bi	209	9.747	0.0281	0.034	9	0.001	0.253
Th	232.05	11.3	0.0293	7.56	12.6	0.222	0.369
U	238.07	18.9	0.04783	7.68	8.3	0.367	0.397
UO_2	270.07	10	0.0223*	7.7	16.7	0.17	0.372
U233				580			4.18
U-235	235.1175	18.7	0.0479	683	15	32.9	0.72
U-238	238.1252	18.9		$2.71\ (n, \gamma)$	8.3		0.478
Pu-239		19.7	0.0497	1029	9.6	51.1	0.478

* Molecules/c.c.

promptly from the fission fragments within 10^{-14} sec of the fission process. By defining

v = average number of neutrons produced per fission, and

η = average number of fission neutrons produced for every neutron *absorbed* in the fuel,

one has:

$$(\Sigma_f^{235} + \Sigma_f^{238})v = (\Sigma_a^{235} + \Sigma_a^{238})\eta$$

in which $\Sigma_f^{238} = 0$ for thermal neutrons. Values of v and η are given in Table 3.2.

TABLE 3.2   Values of v and η.

	Incident neutron energy	U^{235}	Pu^{239}	U^{233}	Natural uranium
v	thermal	2.44	2.90	2.50	—
	1.8 Mev	2.72	3.28	2.75	—
	14.1 Mev	4.52	4.85	3.86	—
η	thermal	2.06	2.10	2.27	1.33
	fast	2.18	2.74	2.60	1.09

It is essential for a fuel to have high enough v and η to account for the absorption and leakage of the neutrons from a reactor and to insure a sufficiently high neutron economy to maintain a chain reaction or to possibily breed new fuels through transmutations at the expense of extra neutrons. The power density of a power reactor at any time is determined by using the neutron concentration and not the gamma concentration, as will be discussed later.

E.   Delayed Neutrons

Delayed neutrons are important in the control of a reactor (with the exception of certain research reactors). By plotting the delayed neutron concentration as a function of time after a reactor is shut down on a semi-log paper and using the technique of graphical stripping to substract off the straight line segments successively, it is possible to identify the half-lives and relative concentrations of different groups of delayed neutrons from the slopes and the intersections of those line segments with the y-axis. There appear to be six groups of delayed neutrons and Table 3.3 is a list of their properties.

TABLE 3.3 Properties of delayed neutrons.

Group	Energy (Mev)	Half-life (sec)*	Number of delayed neutron per thermal fission			Number of delayed neutron per fast fission		
			U-235	Pu-239	U-233	U-235	Pu-239	U-233
1	0.25	55.72	0.00052	0.00021	0.00057	0.00063	0.00024	0.00060
2	0.46	22.72	0.00346	0.00182	0.00197	0.00351	0.00176	0.00192
3	0.41	6.22	0.00310	0.00129	0.00166	0.00310	0.00136	0.00159
4	0.45	2.30	0.00624	0.00199	0.00184	0.00672	0.00207	0.00222
5	0.41	0.610	0.00182	0.00052	0.00034	0.00211	0.00065	0.00051
6	—	0.230	0.00066	0.00027	0.00022	0.00043	0.00022	0.00016
		Total	0.01580	0.00610	0.00660	0.01650	0.00630	0.00700

* For fast or thermal fission of U^{235}, Pu^{239}, or U^{233}, the half-lives are all very close to the indicated values for U-235 (thermal).

D

For thermal fission of U^{235}, for example, since 2.44 neutrons are produced per fission, the ratio of delayed neutrons and total neutrons, β, is thus $0.01580/2.44 = 0.65\%$. Table 3.4 gives the delayed neutron fraction β for various fuels.

TABLE 3.4   Delayed neutron fraction, β.

		β
Thermal fission*	U^{235}	0.0065
	Pu^{239}	0.0021
	U^{233}	0.0026
Fast fission	U^{238}	0.0157
	Th^{232}	0.0220

* The β's for fast fission of U^{235}, Pu^{239}, and U^{233} are approximately that of thermal fission.

It has been shown that the first group of delayed neutrons are due to beta decays of fission fragments $_{35}Br^{87}$ into excited $_{36}Kr^{87}$ and the latter emit neutrons to change to stable $_{36}Kr^{86}$. The fact that $_{36}Kr^{87}$ has 51 neutrons, one more than the "magic number" of 50 for closed-shell stable nucleus, indicates that the last neutron is loosely bound and is easily emitted. Stable Sr^{87} is the end product of Br^{87} through a different decay scheme involving unexcited but radioactive Kr^{87}. The second delayed neutron group is identified with I^{137} and the other four groups are believed to be associated with Br^{89-91}, I^{139}, Sb^{137} or As^{85}, and Li^9, respectively.

F.   Prompt Gammas

Prompt gammas are gammas emitted within 10^{-7} sec of the fission process. The average energy per gamma is roughly 1 Mev. The 7 Mev/fission of prompt gamma energy has to be considered in shielding, heat generation, and gamma-induced nuclear reactions.

G.   Delayed Gammas and Betas

The approximately 6 Mev/fission of delayed gamma energy and 7 Mev/fission of delayed beta energy from the radioactive decays of the fission fragments are important in reactor operation. Even after a reactor is shut down, deliberately or accidentally, these delayed gammas and betas continue to generate heat and steps must be taken to remove this heat so that the fuel elements will not get too hot and suffer damage. The reprocessing of spent fuel is also made more difficult because of these delayed radiation.

If the power level in a reactor is suddenly decreased, the gamma flux does not decrease proportionally, due to the existence of a large fraction of delayed gammas. For this reason, the gamma flux is not used to control the power level of a reactor.

The mean energies of the beta particles and delayed gammas from the fission products are about 0.4 Mev and 0.7 Mev, respectively. From about 10 seconds to about 100 days after fission has occurred, the following formulae by Wigner and Way are accurate to within a factor of two:

Beta emission rate = $3.8 \times 10^{-6} \ t^{-1.2}$ beta/sec-fission (t in days)
Gamma emission rate = $1.9 \times 10^{-6} \ t^{-1.2}$ gamma/sec-fission (t in days)

Total beta and gamma energy emission rate = $2.8 \times 10^{-6} \ t^{-1.2}$ Mev/sec-fission (t in days).

If a reactor is started up at $t = 0$, operated at a constant power of P_0 watts, and shut down at $t = t_0$, then at $t = \tau$ ($\tau > t_0$ and $\tau - t_0$ is the cooling time in days after shutdown) the decay power $P(\tau)$ due to both gammas and betas is obviously:

$$P(\tau)(\text{watts}) = \int_{t_0}^{0} 2.8 \times 10^{-6}(\tau - T)^{-1.2} \ \frac{\text{Mev}}{\text{sec-fis}} \ P_0(\text{watts}) \ dT(\text{days})$$

$$\times \ 3.3 \times 10^{10} \ \frac{\text{fis}}{\text{watt-sec}} \times 1.6 \times 10^{-13} \ \frac{\text{watt-sec}}{\text{Mev}}$$

$$\times \ 8.64 \times 10^{4} \ \frac{\text{sec}}{\text{day}}$$

or,

$$\frac{P(\tau)}{P_0} = 6.5 \times 10^{-3}[(\tau - t_0)^{-0.2} - \tau^{-0.2}] \tag{3-1}$$

where T is an arbitrary time during reactor operation and is a dummy variable. Figure 3.12 is a plot of the Borst-Wheeler curves and is an interpretation by Wheeler of the experimental data on energy liberation by gross fission products as a function of the time of irradiation and decay.

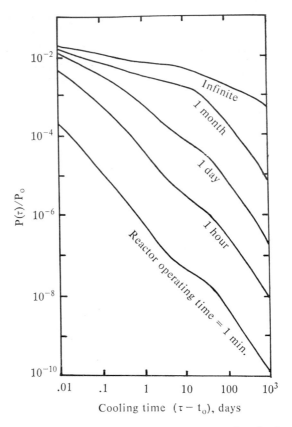

Figure 3.12 Decay power as function of time after shutdown.

H. Fission Cross-section and Non-1/v Factor

Neutron cross-sections (absorption, fission, scattering, etc.) of most materials of interest are given in the well known "barn books" by D. J. Hughes and J. A. Harvey, *i.e.*, "Neutron Cross Sections, BNL-325, and Heavy Element Cross Sections Presented at Geneva, Addendum to BNL-325, U.S. Government Printing Office, Washington, D.C., 1955". Figures 3.7 and 3.9 are samples of what may be found in the barn books.

We have seen (equation 2-2) that for an $1/v$ absorber such as copper and boron, $\bar{\sigma} = (v_p/\bar{v})\sigma_p$, and if the neutron distribution is Maxwellian also, then $\bar{\sigma}(T_0) = \sigma_p(T_0)/1.128$ and $\bar{\sigma}(T) = (T_0/T)^{\frac{1}{2}}\sigma_p(T_0)/1.128$ where T_0 may be

293°K and T is a temperature of interest. However, for certain reactor materials, an empirical correction factor $(g + rs)$ is needed for the expression above to correct for the practical cases of deviations from the Maxwellian neutron flux distribution and for deviations from $1/v$ absorption due to resonances near thermal energy. If the distribution is Maxwellian, $r = 0$. For an $1/v$ absorber in a non-Maxwellian flux, $s = 0$. For a non-$1/v$ absorber in a Maxwellian neutron flux at temperature T, therefore,

$$\bar{\sigma}(T) = g(T) \frac{\sigma(T_o)}{1.128} \left(\frac{T_o}{T}\right)^{1/2} \tag{3-2}$$

Figure 3.10 is a plot of $g(T)$ and Table 3.5 below is a list of the $g(T)$ factors at 293°K for some of the elements for which g is not exactly 1.

TABLE 3.5   Non-$1/v$ factor at 20°C

	g
U-235 absorption	0.975
U-235 fission	0.978
U-238 absorption	1.002
natural U absorption	0.984
Pu-239 absorption	1.074
Pu-239 fission	1.053
Natural Cd absorption	1.32
Xe-135 absorption	1.15
In-115 absorption	1.017

SOLVED PROBLEMS

3.1   A thermal reactor has a fuel load of 75 kg of U-235 and a heat output of 231 Mw. What is the average thermal neutron flux?
Answer: The power of a thermal reactor is related to the neutron flux by

$$P = \frac{V\bar{\Sigma}_f \bar{\phi}}{3.3 \times 10^{10}}$$

Since $P = 231 \times 10^6$ watts, $\bar{\phi}$ is the thermal flux, and

$$V\bar{\Sigma}_f = (M/\rho)(\rho A_0/A)\bar{\sigma}_f$$

where $\bar{\sigma}_f = \dfrac{580 \times 10^{-24}}{1.128} \times 0.978,$ $A_0 = 0.6023 \times 10^{24}$

$M = 75 \times 10^3,$ $A = 235$

We have $\bar{\phi} = 7.87 \times 10^{13}$ neutrons/cm²-sec.

(*Note:* The fuel load and heat output given are that of the Shippingport Pressurized Water Reactor, PWR)

3-2 20 grams of UO_2, the U being 60% enriched, is subjected to a Maxwellian flux ϕ of 10^5 neutrons/cm²-sec at 20°C. What is the approximate fission neutron production rate? Use Table 3.1.

Answer: There are approximately 0.0223×10^{24} UO_2 molecules/c.c. Thus,

$N^{235} = (0.0223 \times 10^{24})(60\%) = 0.0134 \times 10^{24} U^{235}$ atoms/c.c.

$\Sigma_f^{235} = 0.0134 \times 577 = 7.72$ cm⁻¹

From equation 3-2, $\bar{\Sigma}_f = (7.72/1.128)(0.978) = 6.7$ cm⁻¹

The fission rate $R = \phi \bar{\Sigma}_f = 6.7 \times 10^5$ fissions/sec-c.c.

The 20 grams of UO_2 occupies 2 c.c. ($\rho = 10$ g/c.c.). Thus,

$R = 13.4 \times 10^5$ fissions/sec.

The fission neutron production rate $= (2.44$ neutrons/fission$)R$

$$= 3.27 \times 10^6 \text{ neutrons/sec.}$$

3-3 Keeping in mind that Pu^{239} and U^{233} can be "bred" from U^{238} and Th^{232}, respectively, what can be said about fuel production inside a reactor, based on the values of η?

Answer: The value of η for Pu^{239} is 2.74 for fast neutrons. If one of these neutrons is used to breed another U^{238} atom and another one of the 2.74 neutrons is used to sustain a chain reaction, a balance of 0.74 neutron will be available for absorption, leakage, and thermalization, making it likely (and, in fact, possible) to generate power and at the same time to produce more fuel (Pu^{239}) than the fuel used (U^{238}) by means of a fast breeder reactor in which neutrons are kept at high energies with least possible slowing down processes. The low value of 2.10 for thermal neutrons make a Pu^{239} thermal breeder reactor unlikely. However, the thermal value of 2.27 for U^{233} makes Th^{232} a more favorable choice over U^{238} as a thermal breeder reactor fuel.

3-4 It is known that Ba^{137} is a stable final product of the second delayed neutron group. Construct the decay scheme for that group of delayed neutron emitters, using the chart of the nuclides.

Answer:

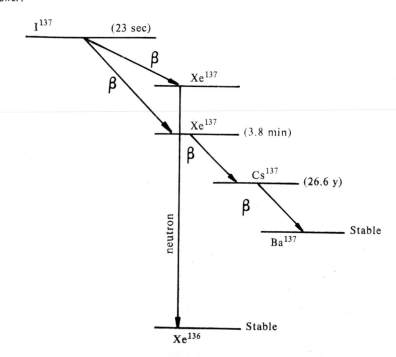

3-5 It is planned to start repair work on a reactor on Friday at 8 a.m. In order to avoid exposure of workers, the reactor is to be run only from 3 p.m. to 4 p.m. the preceding day. How much higher would the gamma level be if the reactor were instead operated from 9 a.m. to 4 p.m. at the same power level?

Answer:

$$\frac{P_2}{P_1} = \frac{(\tau_2 - t_2)^{-0.2} - \tau_2^{-0.2}}{(\tau_1 - t_1)^{-0.2} - \tau_1^{-0.2}}$$

where τ_2 = 9 a.m. to 8 a.m. = 23 hours

τ_1 = 3 p.m. to 8 a.m. = 17 hours

t_2 = 9 a.m. to 4 p.m. = 7 hours

t_1 = 3 p.m. to 4 p.m. = 1 hour

$$\therefore \quad \frac{P_2}{P_1} = \frac{16^{-0.2} - 23^{-0.2}}{16^{-0.2} - 17^{-0.2}} = 5.7$$

3-6 What is the mean-life of the delayed neutrons from the thermal fission of U^{235}?
Answer: The mean life τ (= average length of time for decay to occur) is given by $\tau = (1/\lambda) = T_{\frac{1}{2}}/0.6931$. The delayed neutron fraction for group i, β_i, is the ratio of the number of delayed neutron per thermal fission and ν (= 2.44).
Thus, in table form,

Group i	τ_i	β_i
1	80.55	0.00021
2	32.80	0.00142
3	8.98	0.00127
4	3.32	0.00256
5	0.881	0.00075
6	0.332	0.00027
		$\beta = 0.0065$

$$\bar{\tau} = \frac{\sum_i \beta_i \tau_i}{\sum_i \beta_i} = \frac{0.084}{0.0065} = 12.9 \text{ sec}$$

3-7 What is the average microscopic fission cross-section of a sample of Pu^{239} exposed to a Maxwellian neutron distribution at $400°C$?
Answer: From Figure 3.10, $g(T) = 1.69$.

$$\bar{\sigma}(673°K) = 1.69 \frac{742}{1.128} \left(\frac{273}{673}\right)^{1/2} = 707 \text{ barns}$$

SUPPLEMENTARY PROBLEMS

3-a The binding energy per nucleon for U^{235} is about 7.6 Mev. The average binding energy of the fission fragments is about 8.5 Mev per nucleon (See, also, Problem 1-b). What conclusion on fission energy can you draw from these statements?

3-b Construct a decay scheme for the first delayed neutron group, similar to that of Problem 3-3.

3-c What is the average energy of the first five groups of delayed neutrons from the thermal fission of U^{235}?

3-d A reactor is operated at a constant power level of 500 Mw for one day and is then shut down. What is the gamma and beta heat generation rate 12 hours after shut down? Compare your result with Figure 3.12.

NUCLEAR REACTOR PLANT—GENERAL CONSIDERATIONS

THE FORMULATION of the mass-energy equation by Einstein in 1905, the discovery of nuclear fission by Hahn and Strassman in 1939, and the discovery of liberation of neutrons by fissioning process by Curie-Joliot in 1939 led to the atomic bombs, the nuclear reactors, and the sciences and arts of nuclear engineering. It is the purpose of this chapter to give the readers a general survey of the subject of nuclear engineering and to orient the various topics with regard to the over-all structure and scope of the subject.

A. Nuclear Engineering

Nuclear engineering is the branch of engineering directly concerned with the release, control, and use of all types of energy from nuclear sources. It includes the design and development of systems, such as fission reactors, fusion reactors, and radioisotope energy sources, for the controlled release of nuclear energy and the applications of radiation. It is recognized that nuclear engineering is closely associated not only with the physics and chemistry of nuclear materials, but also with branches of engineering such as chemical, civil, electrical, and mechanical. The nuclear engineer will have the opportunities to concentrate on such areas as interaction of radiation with matter, instrumentation and control, reactor analysis, fuel management, energy transfer, computer applications, economics, space vehicles, radio-isotope applications, legal processes, materials, manufacturing and sales, safety, construction, and administration.

B. Reactor Power Plant

A nuclear reactor power plant may have the following major facilities:

1. *Reactor Building*

a) *Reactor*

    1. Core—fuel, moderator and coolant, control rods, instrumentations, pressure vessel, structural materials, etc., designed to control chain reaction and heat production and removal.

2. Reflector—Designed to scatter neutrons back into the core and reduce leakage. It may contain moderator-type material, supporting structure, coolant, and/or fertile material for breeding.
3. Shielding—A primary shield may be used to shield the gammas and neutrons from the core and vessel. A secondary shield may be used to shield the primary coolant system.

b) *Coolant loops*—Two, three, four, or more primary coolant loops may be associated with each reactor, containing major items such as coolant pumps, steam generators and, in some designs, loop isolation valves.

c) *Service machines*—for fueling, refueling, and in-service inspections.

d) *Safety systems*—Inside the reactor building may be such safety systems as pressure suppression system for the containment, emergency core cooling system (ECCS), and containment cooling systems. These are discussed later in this chapter.

e) *Internal structure*—for missile protection, shielding, equipment support, and pressure transient protection in the event of a major accident inside the containment.

f) *Containment*—to prevent escape of radioactive materials in case of accident.

2. *Electrical Power Conversion Facility*

Secondary coolant loop with relief and safety valves, water purification system, feedwater system, condenser, pump, turbine, generator, and switch-yard.

3. *Auxiliary Building*

This building houses the majority of the radwaste system, fuel storage and transfer facility, engineered safety feature components, etc.

4. *Control Building*

Control room, administration, vital batteries, transformer rooms, cable spreading rooms, etc.

5. *Others*

Diesel generator rooms for emergency power, pumping station and intake building, cooling towers, health and safety offices, etc.

C. Specifications of Nuclear Reactors

There are many ways of classifying nuclear reactors. The following considerations are useful in classifying and specifying reactors:

1. *Product*
   a) Heat—power reactor
   b) Isotopes—high flux reactor
   c) Fuel—breeder reactor
   d) Neutrons and radiations—research and education reactor
   e) Data—research and development reactor

2. *Fuel and Moderator Arrangement*
   a) Heterogeneous
   b) Homogeneous

3. *Energy Spectrum of Neutrons that Produce Fission*
   a) Fast—average energy at about 100 kev
   b) Intermediate—average energy at about 100 ev
   c) Thermal—average energy at about 0.028 ev

4. *Reactor Materials*
   a) Fuel—natural uranium, enriched uranium, Pu-239, U-233.
   b) Coolant—pressurized water, boiling water, D_2O, Na, NaK, He, air, N_2, CO_2, Bi, Hg, etc.
   c) Moderator and reflector—graphite, H_2O, D_2O, Be, BeC, Beryllia (BeO), etc.
   d) Control rod—cadmium, boron carbide, rods containing boron, hafnium, gadolinium, europium, and other rare earths, etc.
   f) Shield—concrete, iron and steel, cadmium and boron, lead and iron, water, etc.
   g) Structure—stainless steel, aluminum, zirconium, beryllium, etc.

D. Reactor Analysis

With the exception of the reactor core, most of a nuclear reactor plant is quite similar to that of a convention fossil steam plant. (There are, of course, problems such as radioactive fuel and coolant handling that a fossil steam plant does not have to consider.) The efforts of most nuclear engineers are, therefore, concentrated on the reactor core. Major reactor physics problems are:

1. Critical-mass calculations
2. Neutron distributions (spatial and energy)
3. Reactor kinetics and stability
4. Fuel burn-up, poison product, and reactor control
5. Accident prevention

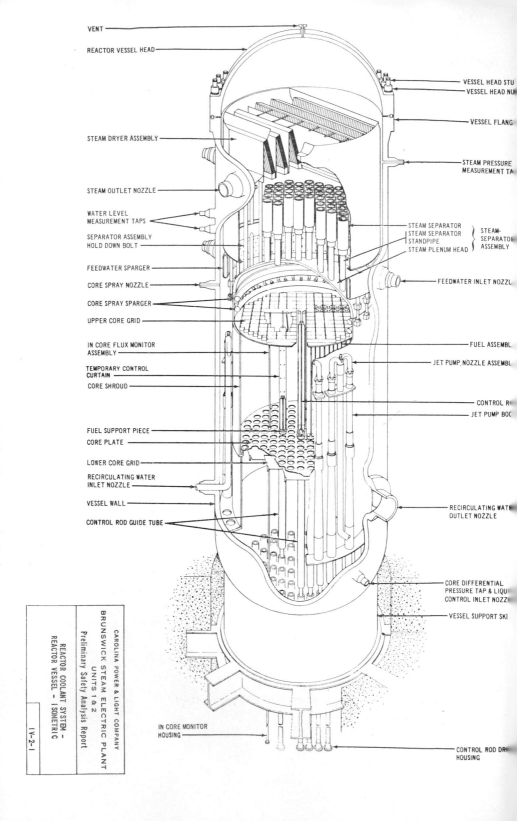

VENT

REACTOR VESSEL HEAD

VESSEL HEAD STU

VESSEL HEAD NU

VESSEL FLANG

STEAM DRYER ASSEMBLY

STEAM PRESSURE
MEASUREMENT TA

STEAM OUTLET NOZZLE

WATER LEVEL
MEASUREMENT TAPS

STEAM SEPARATOR
STEAM SEPARATOR
STANDPIPE
STEAM PLENUM HEAD

STEAM-
SEPARATOR
ASSEMBLY

SEPARATOR ASSEMBLY
HOLD DOWN BOLT

FEEDWATER SPARGER

FEEDWATER INLET NOZZL

CORE SPRAY NOZZLE

CORE SPRAY SPARGER

UPPER CORE GRID

IN CORE FLUX MONITOR
ASSEMBLY

FUEL ASSEMBL

JET PUMP NOZZLE ASSEMBL

TEMPORARY CONTROL
CURTAIN

CORE SHROUD

CONTROL R

JET PUMP BOD

FUEL SUPPORT PIECE

CORE PLATE

LOWER CORE GRID

RECIRCULATING WATER
INLET NOZZLE

VESSEL WALL

RECIRCULATING WATI
OUTLET NOZZLE

CONTROL ROD GUIDE TUBE

CORE DIFFERENTIAL
PRESSURE TAP & LIQUI
CONTROL INLET NOZZ

VESSEL SUPPORT SKI

IN CORE MONITOR
HOUSING

CONTROL ROD DR
HOUSING

CAROLINA POWER & LIGHT COMPANY
BRUNSWICK STEAM ELECTRIC PLANT
UNITS 1 & 2
Preliminary Safety Analysis Report

REACTOR COOLANT SYSTEM -
REACTOR VESSEL - ISOMETRIC

IV-2-1

For most theoretical analysis, the analytical models used are determined by the degree of accuracy wanted and the effort that one is willing to invest. The most useful models and methods are, in increasing accuracies and difficulties:

1. One speed diffusion theory—only spatial distribution is studied.
2. Age-diffusion theory—neutron slowing-down is taken into account also.
3. Multigroup diffusion theory—both spatial and energy distributions can be obtained.
4. Transport theory—the directions of travel of the neutrons are taken into account as well as their spatial and energy distributions.

Techniques such as perturbation theory and numerical analysis using computers are often used in conjunction with the models listed above.

E. Reactor Characteristics and Design Data

Different reactors have, naturally, different characteristics. As an illustration, certain design data and characteristics of the Brunswick Steam Electric Plant—Units 1 & 2 of the Carolina Power and Light Company are given below in Table 4.1. (Courtesy: CP & L and GE). The boiling water reactors are scheduled for completion in 1974 and 1976.

F. Engineered Safety Features

There must be sufficient redundancy of components and power sources in every reactor plant to prevent undue risk to public health and safety under the conditions of any reasonable, hypothetical accident. Some of the major plant abnormalities and transients considered by the design engineers are:

1. Loss of reactor coolant flow due to a rupture in the coolant loop or for any other reason.
2. Loss of all normal AC power to the plant.
3. Loss of external electrical load or excessive load increase.
4. Uncontrolled control rod withdrawal, drop, or malpositioning.
5. Excess feedwater incident.
6. Turbine generator overspeed.
7. Chemical control system malfunction.
8. Startup of an inactive reactor coolant loop.
9. Unexpected turbine trips.
10. Fuel handling accidents.
11. Accidental release of radioactive liquid or waste gas.
12. Rupture of a control rod drive mechanism housing.

TABLE 4.1   Principal plant design features of the Brunswick nuclear plants

1. *SITE*
   Location   Brunswick County,
                                               North Carolina
   Size of Site (acres)   1,200
   Site Ownership   Carolina Power &
                                               Light Company
   Plant Ownership   Carolina Power &
                                               Light Company
   Number of Units on Site   2

2. *PLANT* (per Unit)
   Reactor Rated
      Net Electrical Output (MWe)   821
      Gross Electrical Output (MWe)   847
      Plant Net Heat Rate (Btu/kw-hr)   10,120
      Plant Gross Heat Rate (Btu/kw-hr)   9,816
      Feedwater Temperature (F)   420

   Turbine Valves Wide Open
      Net Electrical Output (MWe)   855
      Gross Electrical Output (MWe)   880
      Plant Net Heat Rate (Btu/kw-hr)   10,119
      Plant Gross Heat Rate (Btu/kw-hr)   9,827
      Feedwater Temperature (F)   424.5

3. *REACTOR VESSEL* (per Unit)
   Inside Diameter (ft-inch)   18-2
   Overall Length Inside (ft-inch)   69-4
   Design Pressure (psig)   1250
   Wall Thickness (inches)   5-17/32

4. *REACTOR COOLANT RECIRCULATION LOOPS* (per Unit)
   Location of Recirculation Loops   —
   Number of Recirculation Loops   2
   Pipe Size (inches)   28
   Pump Capacity, each (gpm)   45,200
   Number of Jet Pumps   20
   Location of Jet Pumps   Inside Reactor Vessel

5. *REACTOR* (per Unit)
   Reactor Rated
      Thermal Output (MWt)   2,436
      Reactor Operating Pressure (psig)   1,005
      Total Reactor Core Flow Rate (lb/hr) 75.5×10^6
      Main Steam Flow Rate (lb/hr) 10.48×10^6
   Turbine Flow Valves Wide Open
      Thermal Output (MWt)   2,550

Reactor Core Description

Lattice	7×7
Control Rod Pitch (inches)	12.0
Number of Fuel Assemblies	560
Number of Control Rods	137
Number of Temporary Control Curtains	248
Effective Active Fuel Length (inches)	144.0
Number of Instrument Tubes	43
Equivalent Reactor Core Diameter (inches)	160.2
Circumscribed Reactor Core Diameter (inches)	170.5
Total Weight UO_2 (lb)	272,850
Total Weight U (lb)	240,512

Reactor Fuel Description

Fuel Material	UO_2
Fuel Density (lb/ft³) @ 93.3% Theoretical	639
Fuel Pellet Diameter (inches)	0.487
Fuel Rod Cladding Material	Zircaloy-2
Fuel Rod Cladding Thickness (inches)	0.032
Fuel Rod Cladding Process	Free Standing, Loaded Tubes
Fuel Rod Outside Diameter (inches)	0.563
Length of Gas Plenum (inches)	16.0
Fuel Rod Pitch (inch)	0.738
UO_2 Weight per Assembly (lb)	487.2
U Weight per Assembly (lb)	429.46
Fuel Assembly Channel Material	Zircaloy-4

Reactor Control (Control Rods)

Number	137
Shape	Cruciform
Material	B_4C Granules Compacted in SS Tubes
Pitch (inches)	12.0
Control Length (inches)	143.0
Blade Span (inches)	9.75
Number of Control Material Tubes per Rod	84
Tube Dimensions (inches)	0.188 O.D. $\times$ 0.025 wall
Stroke (inches)	144.0

Reactor Control (Temporary Control Curtains)

Number	248
Shape	Flat Sheet
Material	Natural Boron-SS
Average Concentration (ppm B)	Approx. 5400
Axial Distribution (from top)	None
Dimensions (inches)	$8.50 \times 0.063 \times 141.25$

Thermal Hydraulic Data

Heat Transfer Area per Assembly (ft²)	36.519
Reactor Core Heat Transfer Area (ft²)	48,451

TABLE 4.1 *cont.*

Maximum Heat Flux (Btu/hr-ft^2)	428,967
Average Heat Flux (Btu/hr-ft^2)	164,743
Thermal Design Limit	
Maximum Power per Fuel Rod Unit Length (kw/ft)	18.50
Average Power per Fuel Rod Unit Length (kw/ft)	7.1
Maximum Fuel Temperature (F)	4380
Minimum Critical Heat Flux Ratio	>1.9
Total Heat Generated in Fuel (%)	96.0
Inlet Subcooling (Btu/lb)	23.9
Core Average Exit Quality	14.1
Power Distribution—Peaking Factors (Peak/Average)	
Axial	1.50
Relative Assembly	1.40
Local (within assembly)	1.24
Overall Total	2.60
Nuclear Design Data	
Average Discharge Exposure (first core)	19,000 MWD/Short Ton U
Initial Enrichment, Average Over Each Assembly	2.25 w/o
Moderator to Fuel Volume Ratio at Total Core	
H_2O/UO_2 Cold	2.41
In-Core Instrumentation	
Number of Power Range (in-core)	
Monitoring Assemblies (fixed)	31
Number of Intermediate Range	
Monitoring Chambers	8
Number of Start-up Range Monitoring Counters	4
Number of Start-up Sources	5
Number of Reactor Vessel Penetrations	43
Reactivity Control	
Reactivity of Core with All Control Rods In	$<0.96\,k_{\text{eff}}$
Reactivity of Core with Strongest Control Rod Out	$<0.99\,k_{\text{eff}}$
Typical Moderator Temperature Coefficient ($\Delta k/k/F$)	
Cold	-5.0×10^{-5}
Hot (no voids)	-17.0×10^{-5}
Operating	—
Typical Moderator Void Coefficient ($\Delta k/k\,\%$ void)	
Cold	-0.55×10^{-3}
Hot (no voids)	-1.0×10^{-3}
Operating	-1.6×10^{-3}
Typical Fuel Temperature (Doppler) Coefficient	
Cold	-1.3×10^{-5}
Hot (no voids)	-1.2×10^{-5}
Operating	$\leq 1.3 \times 10^{-5}$

6. *CONTAINMENT SYSTEMS*
   Primary Containment (per Unit)

Type	Pressure Suppression

Construction

Drywell	Conical & Cylindrical Steel-Lined Concrete Vessel
Pressure Suppression Chamber	Torus/Steel-Lined Concrete Vessel
Pressure Suppression Chamber-Internal Design Pressure (psig)	+62
Pressure Suppression Chamber-External Design Pressure (psi)	+2
Drywell-Internal Design Pressure (psig)	+62
Drywell-External Design Pressure (psi)	+2
Drywell Free Volume (ft³)	164,100
Pressure Suppression Chamber Free Volume (ft³)	124,000
Pressure Suppression Pool Water Volume (ft³)	87,600
Submergence of Vent Pipe below Pressure Pool Surface (ft)	4
Design Temperature of Drywell (F)	300
Design Temperature of Pressure Suppression Chamber (F)	220
Downcomer Vent Pressure Loss Factor	6.21
Break Area/Total Vent Area	0.02
Drywell Free Volume/Pressure Suppression Chamber Free Volume	1.32
Reactor Coolant System Volume/Pressure Suppression Pool Volume	0.214
Drywell Free Volume/Reactor Coolant System Volume	8.8
Calculated Maximal Pressure After Blowdown with no Prepurge	
Drywell (psig)	46
Pressure Suppression Chamber (psig)	28
Initial Pressure Suppression Chamber Temperature Rise (F)	< 50
Leakage Rate (% Free Volume/Day) at Design Pressure	0.5
Secondary Containment (per Unit)	
Type	Controlled Leakage Elevated Release
Construction	
Lower Levels	Reinforced Concrete
Upper Levels	Steel Superstructure and Siding
Roof	Steel Sheeting
Internal Design Pressure (psig)	0.25
Design In-leakage Rate (% Free Volume/Day at 0.25 inches H₂O)	100

E

TABLE 4.1—*cont.*

Elevated Release Point	
Type	Stack
Construction	Reinforced Concrete
Height (above ground)	100 meters

7. *AUXILIARY SYSTEMS*

Core Standby Cooling System (per Unit)	
Core Spray Cooling System	1
High Pressure Coolant Injection System	1
Auto-depressurization System	1
Residual Heat Removal System	
Low Pressure Coolant Injection Subsystem	1
Primary Containment Spray/Cooling Subsystem	1
Reactor Shutdown Cooling Subsystem	1
Plant Standby Coolant Supply System	1
Reactor Auxiliary System (Numbers per Unit)	
Spent Fuel Pool Cooling and Demineralizer System	1
Reactor Cleanup Demineralizer Cooling System	1
Reactor Core isolation Cooling System	1

8. *ELECTRICAL POWER SYSTEMS*

Transmission System	
Outgoing Lines	Unit 2
	4–230 kV (1973)
	Unit 1
	3 Additional 230 kV
	(Total 7–230 kV)
Auxiliary Power Systems	
Incoming Lines	Unit 2
	4–230 kV
	Unit 1
	3 Additional 230 kV
	(Total 7–230 kV)
On-site Sources	
Auxiliary Transformers	2
Start-up Transformers	2
Shutdown Transformers	0
Gas Turbines	0
Standby Diesel-Generator Systems	
Number of Diesel-Generators	4
Number of 4160 V Emergency Buses	4
Number of 480 V Emergency Buses	4
Battery Systems	
Number of 125 V or 250 V Batteries	2 per Unit
Number of 125 V or 250 V Buses	2 per Unit

9. *STRUCTURAL DESIGN*
   Seismic Design

Maximum Design (horizontal g)	0.08
Maximum Credible Earthquake (horizontal g)	0.16

   Wind Design

Maximum Sustained (mph)	130
Tornadoes (mph) (Shutdown only)	300

Depending on the types of reactors, the engineered safety systems are different. The boiling water reactors (BWR) of the General Electric Company, for example, use pressure suppression chambers to retain accidental release of radioactive coolant while the pressurized water reactors (PWR) of Westinghouse Electric, Babcock and Wilcox, and Combustion Engineering use large steel containment vessels as fission product barriers. Even for the PWR's in the United States, there are major differences in designs. For example, the Westinghouse Electric Company uses a large quantity of borated ice stored in the ice condenser internal to the steel containment vessel to act as a passive heat sink to absorb the energy released during a loss-of-coolant accident (LOCA). In addition, an emergency core cooling system injects borated water into the reactor coolant loops to cool the core and the containment spray system provides a spray of cool, borated water to the containment atmosphere so that the steam may be condensed after all the ice in the ice condenser has melted. Furthermore, it is interesting to note that even for the same company, the design of a reactor plant may be quite different from another. For example, the Sequoyah Units 1 & 2 of the Tennessee Valley Authority has an annulus space between the containment vessel and the reactor building of each unit. This barrier, together with the emergency gas treatment system, reduces fission product leakage by providing hold-up time and by collecting the leakage from the containment vessel for filtration. Many other less recent Westinghouse PWR's do not have the annulus space.

G. Plant Site Considerations

There are many factors that determine the selection of a plant site. Among the most important are the following.

1. *Population and Water Supply*

It is essential for a power plant to have ample water supply to remove waste heat. Even with added features such as cooling towers, a river or a large reservoir is still necessary. Because of the potential danger of having an accident and because of the strict requirements on the allowable radiation dose rate at the site boundary for the protection of the general public during normal operation, a nuclear power plant site is generally quite large. The population density of the area surrounding the site should be relatively low. For example, the "minimum exclusion distance" may be around 2000 feet and the "low population distance" (LPD) may be around 3 miles. The "population center distance" may be 10 miles away. Needless to say, the site cannot be too far away from the population center because of the cost in distribution.

2. Meteorology

The atmospheric diffusion characteristics of the site, often measured by the so-called "Pasquill classification", is probably the most important meteorological factor. The probability of having a tornado at the site within any one year is also an important factor. Usually, the effects of tornadic winds of 300 mph rotational velocity plus 60 mph translational velocity, and having 3 psi pressure differential in 3 seconds are considered together with certain tornado generated missiles. For design purposes, tornado generated missiles include the following:

a) A 2-inch by 4-inch by 12-foot board weighing 40 pounds/cu.ft., end on at a speed of 300 mph.

b) A cross tie, 7 inches by 9 inches by $8\frac{1}{2}$ feet weighing 50 pounds/cu.ft., end on at 300 mph.

c) An automobile weighing 4000 pounds at a speed of 50 mph, 25 feet off the ground.

d) A steel pipe 2 inches in diameter by 7 feet long end on at 100 mph.

3. Hydrology

The maximum probable flood, due to such factors as earthquake and dam failures, is considered concurrent with high wind velocity that can add many feet of wave runup. Critical equipment inside the plant, such as emergency diesel generator building, must be protected against the flood.

4. Geology

Bedrock is required for direct and indirect anchorage of fuel storage pool, primary coolant loop, and primary containment, etc.

5. Seismology

All safety equipment and many buildings, such as reactor containment, auxiliary building, control building, diesel building, and intake pumping station, are designed for the design basis earthquake (DBE) or safe shutdown earthquake (SSE). Simultaneous horizontal acceleration of 0.09 g and vertical acceleration of 0.06 g (=2/3 of horizontal) is not uncommon for design purposes.

H. Plant Design

Table 4.1 above lists the principal plant design features of a boiling water reactor (BWR). In this section, certain plant design highlights are discussed.

1. Containment

For a BWR, an old General Electric Company design consists of a dry well, shaped like a light bulb, and a suppression chamber, shaped like a doughnut. It is an over-under design, with the dry well over the suppression chamber underneath. The design parameters are given in Table 4.1. A new containment, called Mark III, by General Electric, has just come into the market and is quite different from the old design. The new design still maintains the pressure suppression principle, again using water, but the doughnut shaped chamber is gone and is replaced by an annular shaped ring of water in a cylindrical shaped containment, very much like that of a pressurized water reactor from the outside. The new containment is much larger, with a design pressure much lower than the 62 psig listed in Table 4.1.

Westinghouse Electric Company's pressurized water reactor has the option of using a pressure reduction system, called the ice condenser system. Basically, the containment is divided into three sections. The lower compartment contains the reactor and the primary coolant system. A divider deck separates the lower compartment from the upper compartment, the latter acts as a reservoir for compressed air in the event of a loss of coolant accident (LOCA) in the lower compartment. An ice condenser, consists of a large quantity of ice, stored in ice baskets, links the upper compartment and the lower compartment. The borated ice acts as an energy absorbing and steam condensing agent in the event of a LOCA, thus reducing the overall pressure of the containment. A typical ice-condenser containment has a total free volume of about 1.25 million cubic feet and a design pressure of about 15 psig. The cylinder may be about 115 feet in diameter.

Dry containment is yet another design. There is no quick pressure suppression system. The containment design pressure P is related to the free volume V approximately by

$$P = \frac{1.1 \times E}{3V}$$

where P is in psig, V is in 10^6 ft^3, and E in 10^6 BTU is the energy released into the containment during the rapid blowdown due to a LOCA. The factor 1.1 provides a 10% margin for the design value over the calculated value from complicated computer codes such as the CONTEMPT code. A typical containment may have $P = 50$ psig and $V = 2.7 \times 10^6$ ft^3, with cylindrical diameter roughly 135 feet. A dry containment usually requires a post-tensioned concrete wall of about 3 feet thick, since a free standing steel containment would either be very large or would have a very thick wall. Steel walls thicker than 1.5 inches requires field stress relief and is very difficult and expensive.

The containment, together with all the penetrations, may have to have leak tightness of the order of 0.2% weight per day at design pressure. The leak tightness is tested periodically. Access into the containment is limited during periods of operation.

A secondary containment is often provided, with an air cleanup system provided for the space between the primary and secondary containment.

2. *Engineered Safety Features (ESF)*

The engineered safety features are structures, systems, and components that are designed against specific accidents so that the health and safety of the general public is not endangered. Major accident considerations are the loss of coolant accident (LOCA), refueling accident, steam line break accident, and control rod drop accident. The plant is designed for design basis earthquake (DBE), maximum possible flood (MPF), tornado, missiles, jet force due to pipe rupture, and electrical blackout, etc., some of the incidents may occur concurrently with others. The most important ESF are discussed below. Each ESF subsystem is redundant in design so that a single active failure of any component does not affect the designed function of the subsystem.

a) *Emergency core cooling system (ECCS)* A pressurized water reactor (PWR) will be used for the discussion. Normally, the reactor is operated at around 2200 psig and 580 F. In the event of a hypothetical double ended rupture of the hot leg of a coolant pipe, for example, the rapid blowdown causes the vessel to depressurize. The core would overheat and the cladding would melt unless some sort of core cooling system is in effect. The ECCS has many subsystems. The high pressure injection system begins to inject borated water into the core through the high pressure pump at around 1500 psig. When the vessel pressure drops to 600 psig, the core flooding system begins to inject borated water into the core through check valves in series with nitrogen pressurized water storage tanks inside the containment. When the pressure continues to drop and reaches about 200 psig, the low pressure injection system begins to inject water from the borated water storage tank located outside of the reactor containment. This low pressure injection system will continue for the duration of the accident to remove decay heat.

Recently, the adequacy of the ECCS is subjected to many heated discussions. The AEC has set interim criteria on the ECCS design. The whole issue is not settled, however.

b) *Containment isolation system* A reactor containment has many penetrations. For a PWR, there are personnel locks (two in series, interlocked such that at least one is closed during reactor power operation), equipment hatch (not to be opened during reactor power operation), fuel transfer tubes, main steam and feedwater lines, other process lines, sampling and instrumentation lines, electrical penetrations, ventilation supply and purge penetrations, etc. When a containment isolation signal exists, due to such events as containment high pressure, containment high radiation level, and/or reactor coolant low pressure, those penetrations that are not required for post-accident operation are isolated. In most cases, a process line has one isolation valve inside the containment and one outside, the two valves having independent trains of signals and means of closure.

c) *Containment cooling system* As discussed earlier in this chapter, the pressure suppression chamber and the ice condenser system are systems that help to reduce the pressure (and hence the temperature) of the containment. For all PWR containments, the containment spray system is an ESF containment cooling system. Upon receiving the proper signals, such as containment high pressure signal, the spray pumps (redundant trains) take suction from the borated water storage tank and spray the containment through spray nozzles at the dome of the containment. The cold water absorbs the heat inside the containment. Often, chemicals are added to the spray to remove radioactive iodines in the containment atmosphere. When the water from the borated water storage tank is emptied, the containment spray system, as well as the low pressure injection system, takes suction from the floor sump of the containment and recirculate the water through redundant heat exchangers.

Another containment cooling system, diversified in design to minimize common mode failure, is the containment air cooling system. It consists of large redundant fans that blow the hot air toward cold water coils. The coils carry emergency raw cooling water, another ESF system.

d) *Emergency air treatment system*   For a reactor building that has a secondary containment, an ESF air treatment system is provided. It consists of 100% redundant trains, each equipped with high efficiency particulate filters (HEPA) to remove praticulates and charcoal adsorbers to remove iodines. The fission products or their daughters that are in the form of noble gases are assumed to pass through the air treatment system without filtration. The air treatment system would either maintain the interior of the secondary containment at a negative pressure to minimize outleakage or/and serve to recirculate the air to allow for mixing, holdup, and decay.

A separate emergency air cleanup system is often provided for the auxiliary building for such event as refueling accident.

The efficiency of the HEPA is "DOP" tested and the charcoal Freon tested.

e) *Other ESF systems*   There are other ESF systems. For instance, there is a reactor protection system and there is a spent fuel cooling system that ensures that the spent fuel storage pool is always provided with a heat removing mechanism. A diesel generator system provides emergency on-site power. An essential control air system supplies air for important equipment and instrumentations. An air conditioning system provides a suitable environment for the main control room to be occupied after a design basis accident. For most nuclear plants, two reactor units are constructed nearby and many systems, including ESF systems, are shared by both units. It is important, however, in the design to ensure that an accident in one unit or one train cannot propagate to another unit or train. For many plant layouts, there are zone separation for this reason.

3. *Auxiliary Systems*

Non-ESF systems and structures of a nuclear plant include such items as turbine building of a PWR, electrical switchyard, radwaste system, reactor coolant water treatment system, component cooling water system, service air system, and fire protection system.

Chapter 5

NEUTRON CYCLE

A. Introduction

SINCE fission neutrons are very energetic while the fission cross-sections of the fuels are high at low energies, it is important for a thermal power reactor to slow down (or moderate) the fission neutrons inside the core in order to achieve optimum neutron economy. (In the study of neutron distribution in a reactor, therefore, reactor physicists are necessarily interested not only in ordinary space but also in energy space.) The neutron slowing-down process is complicated by the resonance peaks (see Figures 3.7 and 3.9) that fast neutrons have to go through before they reach thermal energies. Leakage, absorption, and scattering are all contributing factors to the neutron cycle in a thermal reactor, a schematic diagram of which is given in Figure 5.1. The slowing-down, resonance, and diffusion processes will be described separately below. The diffusion equation and age equation will then be brought in in order that two quantities, the buckling and the age, may be introduced. Discussions of the various factors, f, η, ϵ, p, L_f, L_t, etc., will then complete a basic understanding of the neutron cycle.

B. Neutron Slowing-down

A neutron loses energy and slows down by elastic and inelastic collisions with the nuclei of a medium. Inelastic collisions are important only when the masses of the nuclei are high or when the neutron energies are over about 0.1 Mev. Since most collisions are below that energy (see solved problems 5-1 and 5-2.) only elastic collisions, whose cross-section is essentially independent of the neutron energy, will be considered here.

It is easy to show (see solved problem 5-4) that after one elastic collision with a nucleus with mass number A, the kinetic energy E of the neutron is given by

$$\frac{E}{E_0} = \tfrac{1}{2}[(1 + \alpha) + (1 - \alpha) \cos \theta] \tag{5-1}$$

where

$$\alpha = \left(\frac{A - 1}{A + 1}\right)^2$$

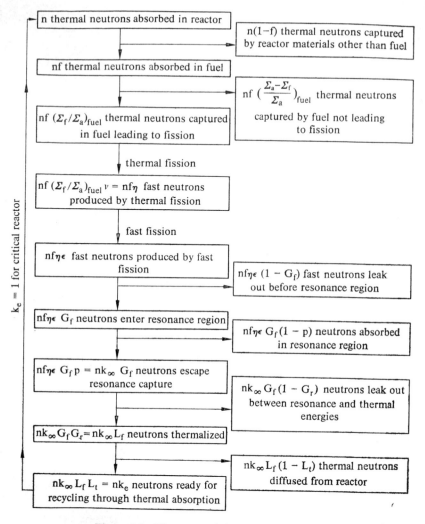

Figure 5.1   Neutron cycle in a thermal reactor.

E_0 = original kinetic energy of the neutron

θ = angle of scattering in the center of mass system

The minimum possible energy after one collision is when $\theta = 0$ (head-on collision) and

$$E_{\min} = \alpha E_0 \tag{5-2}$$

Equation 5-2 provides a physical significance to α. Next, a new convenient quantity u, called the lethargy, is defined as

$$u = \ln (E_0/E) \tag{5-3}$$

It follows that

$$u_2 - u_1 = \ln (E_1/E_2) \tag{5-4}$$

where E_1 and E_2 are two particular energies of interest and u_1 and u_2 are the corresponding lethargies. By defining a quantity ξ as the average of $(u_2 - u_1)$, i.e.,

$$\xi = \overline{u_2 - u_1} \tag{5-5}$$

one can show (see solved problem 5-5) that

$$\xi = 1 + \frac{\alpha}{1 - \alpha} \ln \alpha \tag{5-6}$$

and ξ, the average logarithmic energy decrement per collision or the average lethargy increment per collision, is energy independent. In other words, regardless of its initial energy, a neutron must undergo the same average number of collisions in a particular medium to increase its lethargy by a given amount. This is, of course, why u and ξ are defined in the first place. Illustrative examples are given in solved problems 5-1 and 5-2. It is interesting to note that it takes, on the average, 18 collisions to thermalize a fission neutron in hydrogen, although it is possible for the neutron to lose all its energy in one collision (equation 5-2, $\alpha = 0$ for hydrogen). In passing, it may be noted that for $A \neq 1$,

$$\xi \approx \frac{2}{A + \frac{2}{3}} \tag{5-7}$$

and the fractional error between equations 5-6 and 5-7 is very close to $1/(9A^2)$.

C. Resonance Absorption and Scattering

In a nuclear reaction, when the energy of the incident particle is such that the resulting excited state of the compound nucleus is very close to one of the quantum energy levels of the compound nucleus, there is usually a very marked increase in the reaction rate and is called resonance. In the case that the resonance energy peaks are sufficiently widely separated so that the resonances do not interfere with one another, the resonance cross-section of many reactor materials is represented to a good approximation by the single-level Breit-Wigner formula

$$\sigma(E) = \frac{\sigma_0}{1 + [(E - E_r)/(\Gamma/2)]^2} \tag{5-8}$$

where E is the energy of the neutron, relative to the nucleus, σ_0 is the cross-section at the peak resonance energy E_r and Γ is the total width (in energy) of the resonance peak at half the maximum height (in barns).

In a resonance peak,

$$\sigma = \sigma_{rs} + \sigma_{ra} = \frac{\Gamma_n}{\Gamma} \sigma_0 + \frac{\Gamma_\gamma}{\Gamma} \sigma_0$$

where σ_{rs} and σ_{ra} are the resonance scattering and resonance absorption cross-sections, respectively, and Γ_n/Γ and Γ_γ/Γ are the relative radiative capture (n, γ) and resonance scattering (n, n) rates. The measured total cross-section is

$$\sigma_t = \sigma + \sigma_s$$

where σ_s is due to non-resonance potential scattering. In general, σ_{rs} is small outside a resonance peak and is about the same order of magnitude as σ_s is inside a peak. Thus, it is usually accepted that σ_s is a constant, especially true for $E < 0.1$ Mev and low A.

On the other hand, one can readily appreciate the difficulty involved if and when complicated energy dependent absorption and scattering cross-sections have to be taken into account in the study of neutron slowing-down processes.

D.   Neutron Diffusion and Leakage

Neutrons in a reactor not only slow down (*i.e.*, go from one point to another in energy space) but also diffuse from a region (in ordinary position space) of higher neutron density to a region of lower neutron density. Diffusion can, indeed, take place for monoenergetic neutrons or a collection of neutrons of any energy distribution. For the moment, it will be assumed that the neutrons are monoenergetic. Fick's law of diffusion gives

$$j = -D_c \frac{dn}{dr}$$

$$= -D \frac{d\phi}{dr} \tag{5-9}$$

where

j = neutron current density, (neutrons/cm²-sec)
D_c = conventional neutron diffusion coefficient, (cm²/sec)
D = neutron flux diffusion coefficient, (cm)
n = neutron density, (neutrons/c.c.)
$\phi = nv$ = neutron flux, (neutrons/cm²-sec)
r = a position coordinate, (cm)

and the net neutron leakage through diffusion from an unit volume per unit time, $\mathscr{L}$ (neutrons/c.c.-sec) is given by

$$\mathscr{L} = \text{div } (\mathbf{j}) = \mathbf{V} \cdot \mathbf{j} = \mathbf{V} \cdot (-D\mathbf{V}\phi)$$

or,

$$\mathscr{L} = -D\mathbf{V}^2\phi \tag{5-10}$$

If the medium is a weak absorber, so that $\Sigma_a/\Sigma_t \ll 1$, transport theory (which is beyond the scope of this text) gives

$$D = \cfrac{1}{3\Sigma_s\left(1 - \cfrac{2}{3A}\right)}$$

where A is the atomic mass of the nuclides composing the scattering medium.

E.   Diffusion Equation

It is necessary to know the spatial distribution of the neutrons in a reactor, $n(r)$, in order to calculate the various reaction rates. The position dependent neutron density $n(r)$ is given by the monoenergetic neutron balance equation:

Net Gain = Source − Absorption − Leakage

Each of the four terms above has an unit of (neutrons/c.c.-sec). Thus,

$$\frac{\partial n(r, t)}{\partial t} = S(r, t) - \Sigma_a\phi(r, t) + D\mathbf{V}^2\phi(r, t) \tag{5-11}$$

Equation 5-11 is called the diffusion equation.

At steady state, $\partial n/\partial t = 0$ and

$$D\mathbf{V}^2\phi(r) - \Sigma_a\phi(r) + S(r) = 0 \tag{5-12}$$

In a homogeneous reactor core, in which the fuel is mixed uniformly with the moderating materials, the source of neutrons is fission of the fuel and $S(r)$ is therefore proportional to $\phi(r)$. It is thus obvious that equation 5-12 can be written as

$$\mathbf{V}^2\phi(r) + B^2\phi(r) = 0 \tag{5-13}$$

where B^2 is called the buckling. Solved with the boundary conditions, equation 5-13 will give the shape of the flux distribution in a reactor and determine the value of B^2.

It should be emphasized that the assumptions so far are:
1. monoenergetic neutrons,
2. weak absorption relative to scattering, and
3. the flux gradient does not change rapidly.

The last two assumptions are inherent in the derivation of the leakage term.

F. Age Equation

Having discussed the simple case of diffusion theory, in which monoenergetic neutrons are assumed, attention is now turned to the study of neutron moderation, accompanied by leakage and absorption.

By defining a new quantity, $q(r, E)$, the slowing down density, as the number of neutrons crossing a level of energy E per unit volume per second, the value of $q(r, E)$ at E would be the same as that at $E + dE$ in an infinite, non-absorbing medium. The number of collisions needed for $q(r, E)$ neutrons to cross a differential energy range dE at E is

$$q(r, E) \frac{d(\ln E)}{\xi}$$

where ξ, the average change in $\ln E$ per collision, is defined earlier. However, the number of collisions per unit volume per sec in dE is

$$\phi(r, E)\Sigma_s \, dE$$

Thus, equating,

$$\phi(r, E)\Sigma_s \, dE = q(r, E) \frac{d(\ln E)}{\xi}$$

or,

$$\phi(r, E) = \frac{q(r, E)}{\xi \Sigma_s E} \qquad (5\text{-}14)$$

It shall be assumed that equation 5-14 is accurate even with leakage and weak absorption. A neutron balance equation similar to that of equation 5-12 is now written as

$$D(E)\nabla^2 \phi(r, E) - \Sigma_a(E)\phi(r, E) + S(r, E) = 0 \qquad (5\text{-}15)$$

in which the source term $S(r, E)$ is evidently

$$S(r, E) = \frac{q(r, E + dE) - q(r, E)}{dE} = \frac{\partial q(r, E)}{\partial E} \tag{5-16}$$

Equation 5-15 becomes

$$\frac{D}{\xi \Sigma_s E} \nabla^2 q(r, E) - \frac{\Sigma_a}{\xi \Sigma_s E} q(r, E) + \frac{\partial q(r, E)}{\partial E} = 0 \tag{5-17}$$

Defining a new variable, τ, the age, as

$$\tau = \int_E^{E_0} \frac{D \, dE}{\xi \Sigma_s E} \quad \text{or} \quad d\tau = -\frac{D \, dE}{\xi \Sigma_s E} \tag{5-18}$$

Equation 5-17 becomes

$$\nabla^2 q(r, E) - \frac{\Sigma_a}{D} q(r, E) = \frac{\partial q(r, E)}{\partial \tau} \tag{5-19}$$

where $D = D(E)$, $\Sigma_a = \Sigma_a(E)$, and $\tau = \tau(E)$.

Equation 5-19 is called the age equation. Since the source $S(r, E)$ is related to $q(r, E)$ through equation 5-16, one often has to solve equations 5-15 and 5-19 simultaneously to get $\phi(r, E)$.

The age τ has an unit of cm^2. It is called age because it plays the same role in the age equation as time in a similar differential equation of heat conduction. Solved problem 5-6 will provide a physical interpretation of τ.

G. Thermal Utilization, f

The thermal utilization f for a homogeneous reactor is defined as

$$f = \frac{\text{absorption in fuel}}{\text{absorption in fuel} + \text{absorption in other reactor materials}}$$

$$= \frac{N(235)\sigma_a(235) + N(238)\sigma_a(238)}{N(235)\sigma_a(235) + N(238)\sigma_a(238) + N(m)\sigma_a(m)} \tag{5-20}$$

where uranium fuel has been used as an example and $N(m)$ is the number of nuclei per c.c. of materials other than uranium.

H.   Resonance Escape Probability, p

The resonance escape probability p is the probability that fast neutrons will be moderated past the resonance peaks without resonance absorption. Since the fractional number of neutrons dn/n removed by absorption in crossing a lethargy range du is the product of the number of collisions du/ξ and the relative chance of absorption Σ_a/Σ_t, one has

$$\frac{dn}{n} = - \frac{du}{\xi} \frac{\Sigma_a}{\Sigma_t}$$

Integrating,

$$n = n_0 \exp\left(- \int_{u_0}^{u} \frac{\Sigma_a}{\xi\Sigma_t} du\right) = n_0 \exp\left(- \int_{E}^{E_0} \frac{\Sigma_a}{\xi\Sigma_t} \frac{dE}{E}\right)$$

or,

$$p = \frac{n}{n_0} = \exp\left(- \int_{E}^{E_0} \frac{\Sigma_a}{\xi\Sigma_t} \frac{dE}{E}\right) \tag{5-21}$$

The expression above is generally applicable to homogeneous water-uranium mixtures and to other moderators if the resonance peaks are widely spaced and narrow in comparison with ξ.

I.   Fast Fission Factor, ϵ

The fast fission factor ϵ is the ratio of total fast neutrons produced by fissioning with neutrons of all energies to the number of fast neutrons produced by thermal neutron fissions. If all fission neutrons are monoenergetic with lethargy $u = 0$, it can be shown that

$$\epsilon = \frac{\int_{0}^{u} \gamma(u)p(u)\, du + \Sigma_f^{th} p_{th}/\Sigma_a^{th}}{\Sigma_f^{th} p_{th}/\Sigma_a^{th}} \tag{5-22}$$

where

$$\gamma(u) = \frac{\Sigma_f(u)}{\xi\Sigma_t(u)}$$

$p(u)$ = resonance escape probability from $u = 0$ to u

p_{th} = resonance escape probability from $u = 0$ to u_{th}.

J. Infinite Multiplication Factor, k_∞

The infinite multiplication factor k_∞ is defined by the so-called four-factor formula:

$$k_\infty = \eta f p \epsilon \tag{5-23}$$

and is the ratio of the number of neutrons resulting from fission in one generation to the number absorbed in the preceding generation, provided that no neutron leaks out of the system (*i.e.*, if the system is infinite in size).

K. Fast Non-leakage Probability, L_f

The fast non-leakage probability $L_f(\tau)$ is defined as the probability that a fast source neutron ($\tau = 0$) does *not* leak out of the system during the course of slowing down to age τ. Consider the age equation:

$$\nabla^2 q(r, \tau) - \frac{\Sigma_a(\tau)}{D(\tau)} q(r, \tau) = \frac{\partial q(r, \tau)}{\partial \tau} \tag{5-19A}$$

The slowing-down density at $\tau = 0$ is equal to the production rate by thermal fission. *i.e.*,

$$q(r, 0) = \phi(r) v \epsilon \Sigma_f{}^u$$

where $\Sigma_f{}^u$ is the macroscopic fission cross-section of the fuel. The slowing-down density at τ is evidently given by

$$q(r, \tau) = q(r, 0) L_f(\tau) p(\tau)$$

where

$$p(\tau) = \exp\left(-\int_\tau^0 \frac{\Sigma_a}{D} d\tau\right) \tag{5-21A}$$

in a weak-absorbing medium where $\Sigma_t \approx \Sigma_s$. Substituting the expression above into equation 5-19A, one gets

$$\nabla^2 \phi(r) = \frac{1}{L_f(\tau)} \frac{\partial L_f(\tau)}{\partial \tau} .$$

From the definition of buckling,

$$\nabla^2 \phi(r) + B^2 \phi(r) = 0 \tag{5-13}$$

one has

$$\frac{\partial L_f(\tau)}{\partial \tau} - B^2 L_f(\tau) = 0$$

or,

$$L_f(\tau) = e^{-B^2 \tau} \tag{5-24}$$

F

L.   Thermal Non-leakage Probability, L_t

The thermal non-leakage probability L_t is the fraction of the neutrons thermal-ized in the reactor which are absorbed as thermal neutrons.
i.e.,

$$L_t = \frac{\text{thermal absorption}}{\text{thermal absorption} + \text{thermal leakage}}$$

Now, since

$$\text{leakage} = -D\nabla^2\phi$$

$$= DB^2\phi$$

and

$$\text{absorption} = \Sigma_a\phi$$

one has

$$L_t = \frac{\Sigma_a}{\Sigma_a + DB^2}$$

or,

$$L_t = \frac{1}{1 + L^2B^2} \tag{5-25}$$

where the thermal diffusion length L is defined by

$$L^2 = \frac{D}{\Sigma_a} \tag{5-26}$$

M.   Effective Multiplication Factor, k_e, and Criticality Condition

The neutron cycle has now been completely described. The n original neutrons have been multiplied by a factor

$$k_e = k_\infty L_f L_t = \eta p f \epsilon L_f L_t$$

If this factor k_e, called the effective multiplication factor, is unity, the neutron cycle has not resulted in an increase or decrease of neutron population and the reactor is said to be critical. In other words, the criticality condition is:

$$k_e = \frac{\eta p f \epsilon \, e^{-B^2\tau}}{1 + B^2L^2} = 1 \tag{5-27}$$

For $k_e < 1$, the reactor is said to be subcritical while for $k_e > 1$, the reactor is supercritical.

SOLVED PROBLEMS

5-1 On the average, how many collisions are required to
  a) thermalize a fission neutron in graphite?
  b) slow down a fission neutron to 0.1 Mev?
Answer: For graphite, $A = 12$. For fission neutron, $E_f = 2$ Mev. For thermal neutron, $E_t = 0.025$ ev. From equations 5-6 or 5-7, $\xi = 0.158$.
a) From equation 5-4,

$$u_t - u_f = \ln (E_f/E_t) = 18.2$$

Thus, the average number of collisions, C_t, for thermalization is given by

$$C_t = \frac{\text{total lethargy change}}{\text{average lethargy change}} = \overline{(u_t - u_f)/(u_t - u_f)}$$

$$= (u_t - u_f)/\xi = 18.2/0.158 = 114$$

b) The intermediate energy E_i is 0.1 Mev. From equation 5-4,

$$u_i - u_f = \ln (E_f/E_i) = \ln 20 = 3$$
$$C_i = 3/0.158 = 19$$

In other words, in the process of thermalizing a fission neutron, most collisions occur below 0.1 Mev.

5-2 Tabulate C_i and C_t (see problem above) for H, D, He, Li, Be, C, O, Na, and U.
Answer:

	A	ξ	C_t	C_i
H	1	1.000	18	3
D	2	0.725	25	4.15
He	4	0.425	43	7.1
Be	9	0.209	86	14.4
C	12	0.158	114	19
O	16	0.120	150	25
Na	23	0.0845	216	35.6
U	238	0.00838	2172	358

The ratio of C_i and C_t is the same in all cases.

5-3 The effective ξ for a compound depends on the scattering cross-sections of the elements forming the compound. If the scattering cross-sections of H and O are 20 barns and 3.8 barns, respectively, within the energy range of 1 kev to 1 ev,
  a) what is the effective ξ for H_2O?
  b) what is the average number of collisions needed to slow a 1 kev neutron to 1 ev?
Answer:
a) For H, $\xi_H = 1$. For O, $\xi_0 = 0.122$. For H_2O,

$$\xi = \frac{2\sigma_H \xi_H + \sigma_0 \xi_0}{2\sigma_H + \sigma_0} = 0.924$$

b)
$$C = \frac{\ln\left(\frac{1000}{1}\right)}{0.924} = 7.5$$

5-4 Prove equation 5.1.

Answer: Using the notations given in Figure 5.2 below, the speed v_c of the center of mass

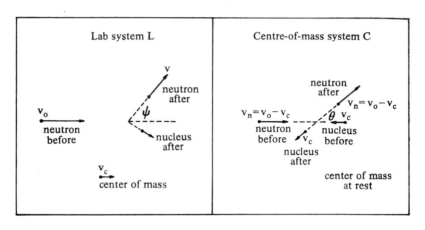

Figure 5.2 Elastic scattering diagram.

relative to an observer in the lab system is given by the requirement that the total momentum of the system of mass $(A + 1)$ and speed v_c must be equal to that of the neutron. Thus,

$$(m + Am)v_c = mv_0$$

or,

$$v_c = \frac{v_0}{1 + A}$$

Also,

$$v_n = v_0 - v_c = \frac{v_0 A}{1 + A}$$

Using the law of cosine (see Figure 5.3),

$$v^2 = v_n^2 + v_c^2 + 2v_n v_c \cos \theta$$

$$= \left(\frac{v_0 A}{1 + A}\right)^2 + \left(\frac{v_0}{1 + A}\right)^2 + 2\left(\frac{v_0 A}{1 + A}\right)\left(\frac{v_0}{1 + A}\right)\cos \theta$$

or,

$$\frac{E}{E_0} = \frac{v^2}{v_0^2} = \frac{A^2 + 2A \cos \theta + 1}{(A+1)^2} = \tfrac{1}{2}[(1 + a) + (1 - a) \cos \theta]$$

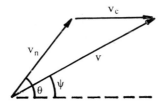

Figure 5.3 *C* system and *L* system.

It should be noted that for $A \gg 1$, the center of mass is essentially stationary in the *L* system and $\theta = \psi$.

5-5 Prove equation 5-6.
Answer: According to equation 5-1, E is a linear function of $\cos \theta$. Since all values of $\cos \theta$ are equally probable, all values of E have to be equally probable. The probability $P\, dE_2$ that a neutron of initial energy E_1 will have energy between E_2 and $(E_2 + dE_2)$ after one collision is

$$P\, dE_2 = \frac{dE_2}{E_1 - E_{\min}} = \frac{dE_2}{E_1 - \alpha E_1}$$

thus,

$$\xi = \overline{\ln \left(\frac{E_1}{E_2}\right)} = \frac{\int_{\alpha E_1}^{E_1} \ln (E_1/E_2) P\, dE_2}{\int_{\alpha E_1}^{E_1} P\, dE_2}$$

$$= 1 + \frac{\alpha}{1 - \alpha} \ln \alpha$$

The derivation is exact. Another way to obtain equation 5-6 is by setting

$$\xi = \frac{\int_{-1}^{1} \ln (E_1/E_2)\, d(\cos \theta)}{\int_{-1}^{1} d(\cos \theta)}$$

where

$$\frac{E_1}{E_2} = \frac{(A + 1)^2}{A^2 + 2A \cos \theta + 1}$$

5-6 Plot $q(r, \tau)$ against r and τ for an unit source of monoenergetic fast neutrons in an infinite homogeneous nonabsorbing medium and prove that $\overline{r^2} = 6\tau$.
Answer: The solution of the age equation

$$\nabla^2 q(r, \tau) = \frac{\partial q(r, \tau)}{\partial \tau}$$

for this case is

$$q(r, \tau) = \frac{\exp\left[-r^2/(4\tau)\right]}{(4\pi\tau)^{3/2}} \tag{5-28}$$

and can be proved by substitution. It is noted that for a given τ, $q(r)$ follows a Gaussian distribution. It is also noted that since τ is a monotonically decreasing function of the energy E, as is evident from equation 5-18, τ is also a monotonically increasing function of the lethargy u.

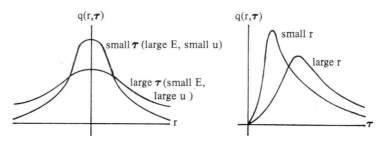

Figure 5.4   Slowing-down density of a point source.

Figure 5-4 is a plot of $q(r, \tau)$.
Substituting $q(r, \tau)$ into the expression

$$\bar{r^2} = \frac{\int r^2 q(r, \tau)\, dV}{\int q(r, \tau)\, dV} = \frac{\int_0^\infty r^2 q(r, \tau) 4\pi r^2\, dr}{\int_0^\infty q(r, \tau) 4\pi r^2\, dr}$$

one has $\bar{r^2} = 6\tau$. This may be used as the expression to attach a physical significance to τ. It may be useful to note from equation 5-28 that for a given τ,

$$\ln q \propto -r^2$$

If $\ln q$ is plotted against r^2, a straight line would be obtained, the slope of which is $-1/(4\tau)$. Experimental determination of τ can be based on this reasoning.

5-7   Go through a light literature search and list the approximate values of ξ, D, τ, L, the slowing-down time t_s, and the diffusion time t_t for H_2O, D_2O, Be, BeO, and graphite. Assume fission neutron energy of 2 Mev.
Answer:

TABLE 5.1   Moderator constants at 20°C

	D (cm)	ξ	$\sqrt{\tau}$ (cm)	L (cm)	M (cm)	t_s (cm)	t_t (sec)
H_2O	0.17	0.93	5.6	2.76	6.2	7.1×10^{-6}	2.4×10^{-4}
D_2O	0.85	0.51	11.0	100	101	5.0×10^{-5}	6.0×10^{-2}
Be	0.54	0.206	9.2	21	23	5.7×10^{-5}	4.2×10^{-3}
BeO	0.66	0.17	10	30	31.6		
C	0.94	0.158	18.7	54.2	57	1.4×10^{-4}	1.6×10^{-2}
He		0.425		2.8×10^5			

where $\sqrt{\tau}$ = slowing-down length (τ = age)

   L = diffusion length

   $M = \sqrt{L^2 + \tau}$ = migration length

Note: a)   No attempt was made in getting the latest accepted values. However, the data given here are definitely sufficient in giving a general idea of the orders of magnitude of the various constants.

   b)   A small amount of H_2O impurity in D_2O can change the values for the latter appreciably.

   c)   The age is given for fission neutrons (2 Mev) to 0.025 ev neutrons. Many literatures provide values for fission neutrons to 1.44 ev neutrons (resonance peak of In-115 activation). As an example, the current accepted age value in H_2O for fission neutrons to 1.44 ev neutrons is 27.6 cm² while one of the latest theoretical values given is 25.9 cm².

   d)   Values of ξ are for epithermal neutrons (see problem 5-3)

   e)   References: "Nuclear Reactor Engineering" by Glasstone and Sesonske, 1963, and "The Theory of Neutron Slowing Down in Nuclear Reactors" by Ferziger and Zweifel, 1966.

SUPPLEMENTARY PROBLEMS

5-a   Prove that the mean square distance $\overline{r_t^2}$ which a thermal neutron travels from its birth (thermalization) to where it is absorbed is given by

$$\overline{r_t^2} = 6L^2$$

(*Hint:* Solve equation 5-12 for $r \neq 0$ and apply the source condition $\lim_{r \to 0} (4\pi r^2 j) = 1$ to prove that

$$\phi = \frac{\exp(-r/L)}{4\pi Dr} \text{ first.})$$

5-b   Suggest an experimental method to measure L.
(*Hint:* Solved problem 5-6).

Chapter 6

REACTOR FLUX DISTRIBUTIONS AND CRITICALITY CONDITIONS

To ILLUSTRATE the basic principles and techniques in reactor analysis, four different representative cases for the neutron flux distributions are described below.

Case A. Bare homogeneous core
One group diffusion method
Steady state

Assuming that all neutrons have the same energy (one group) and starting with the wave equation

$$\nabla^2\phi + B^2\phi = 0 \tag{5-13}$$

where the Laplacian

$$\nabla^2 = \frac{\partial^2}{\partial x^2} + \frac{\partial^2}{\partial y^2} + \frac{\partial^2}{\partial z^2}$$ (rectangular coordinates)

$$\nabla^2 = \frac{\partial^2}{\partial r^2} + \frac{2}{r}\frac{\partial}{\partial r}$$ (spherical coordinates with symmetry with respect to the origin)

$$\nabla^2 = \frac{\partial^2}{\partial r^2} + \frac{1}{r}\frac{\partial}{\partial r} + \frac{\partial^2}{\partial z^2}$$ (cylindrical coordinates with axial symmetry)

one has, in the case of a cylindrical core,

$$\frac{\partial^2\phi(r, z)}{\partial r^2} + \frac{1}{r}\frac{\partial\phi(r, z)}{\partial r} + \frac{\partial^2\phi(r, z)}{\partial z^2} + B^2\phi(r, z) = 0$$

Using the method of separation of variables*, one writes

$$\phi(r, z) = R(r)Z(z)$$

By substitution,

$$\frac{1}{R}\left(\frac{d^2R}{dr^2} + \frac{1}{r}\frac{dR}{dr}\right) + \frac{1}{Z}\frac{d^2Z}{dz^2} + B^2 = 0$$

The variables r and z are evidently separated. Since r and z are independent variables, each of the terms of the equation above may be set equal to a

---

* There is no guarantee that this method works, but when it does, the partial differential equation is reduced to a set of ordinary differential equations and the solution thus obtained is genuine, provided that all boundary conditions are taken into consideration.

constant. Writing*

$$\frac{1}{R}\left(\frac{d^2R}{dr^2} + \frac{1}{r}\frac{dR}{dr}\right) = -\alpha^2 \tag{6-1}$$

$$\frac{1}{Z}\frac{d^2Z}{dz^2} = -\beta^2 \tag{6-2}$$

$$B^2 = \alpha^2 + \beta^2 \tag{6-3}$$

and, subjecting to the boundary condition that the flux must vanish at the surface of the reactor and remain finite inside†, the non-trivial solutions (see solved problem 6-1) of equations 6-1 and 6-2 are

$$R(r) = AJ_0(\alpha r) = AJ_0\left(\frac{2.405r}{R_0}\right)$$

$$Z(z) = C\cos(\beta z) = C\cos\left(\frac{\pi z}{H_0}\right)$$

where A and C are proportional constants, and R_0 and H_0 are the critical radius and height, respectively, with extrapolated distances already taken into account. Thus,

$$\phi(r, z) = \phi_m \cos\left(\frac{\pi z}{H_0}\right) J_0(2.405r/R_0) \tag{6-4}$$

where the origin has been taken at the center of the axis and ϕ_m is the flux at the origin.

The average flux can be obtained by integrating the flux over a differential volume element of $dV = 2\pi r\, dr\, dz$ and then divided by the total volume. Thus,

$$\bar{\phi} = \frac{\int_{z=-H_0/2}^{H_0/2} \int_{r=0}^{R_0} \phi(r,z)2\pi r\, dr\, dz}{\pi R_0{}^2 H_0} = 0.275\phi_m$$

The buckling B^2 is given by $B^2 = \alpha^2 + \beta^2$. This B^2 is a function of geometry only and is called geometric buckling. *i.e.,*

$$B_g{}^2 = \left(\frac{2.405}{R_0}\right)^2 + \left(\frac{\pi}{H_0}\right)^2$$

---

* It will be shown (solved problem 6-1) that both α and β are real and positive.

† Actually, the flux does not vanish at the geometric boundary. Instead it vanishes at an "extrapolated distance" equal to $0.71/[\Sigma_s(1 - \overline{\cos\theta})]$ outside the geometric boundary, according to transport theory. Here, $\overline{\cos\theta}$ is the average cosine of the angle of neutron scattering and is equal to $2/(3A)$ for scattering element with atomic mass A.

The minimum critical volume V_{min} for a given B_g^2 is found by setting $dV/dH = 0$ where

$$V = \pi R^2 H = \frac{\pi(2.405)^2 H^3}{B_g^2 H^2 - \pi^2}$$

This leads to

$$H = \frac{\pi\sqrt{3}}{B_g}, \qquad R = \left(\frac{2.405}{B_g}\right)\sqrt{\frac{3}{2}}, \quad \text{and} \quad V_{min} = \frac{148.2}{B_g^3}.$$

For a critical bare homogeneous reactor of any geometry, the geometric buckling B_g^2 is equal to the so-called material buckling B_m^2.

$$B_g^2 = B_m^2 \quad \text{(critical reactor only)}$$

where B_m^2 is defined from the critical condition

$$\frac{k_\infty e^{-B_m^2 \tau}}{1 + L^2 B_m^2} = 1 \qquad \text{(bare homogeneous core)} \qquad (5\text{-}27)$$

and B_m^2 is a function of the reactor materials only.

Similar analysis for other geometries leads to the results tabulated in Table 6.1. Figure 6.1 provides an indication of the relative flux distributions of the various geometries.

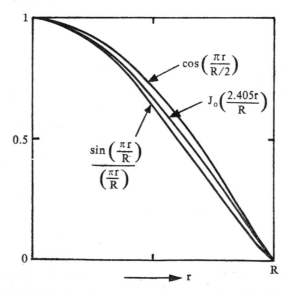

Figure 6.1 Functions for determination of flux distribution

TABLE 6.1 Flux distributions in bare homogeneous reactors.

Core*	B_g^2	ϕ	$\bar{\phi}/\phi_m$	Minimum critical volume
Finite cylinder, height H, radius R	$\left(\frac{2.405}{R}\right)^2 + \left(\frac{\pi}{H}\right)^2$	$\phi_m J_0\left(\frac{2.405r}{R}\right)\cos\left(\frac{\pi z}{H}\right)$	$0.432\left(\frac{2}{\pi}\right)$	$\frac{148}{B^3}$
Sphere, radius R	$\left(\frac{\pi}{R}\right)^2$	$\phi_m\dfrac{\sin\left(\frac{\pi r}{R}\right)}{\frac{\pi r}{R}}$	$\frac{3}{\pi^2}$	$\frac{130}{B^3}$
Rectangular parallelepiped sides a, b, c	$\left(\frac{\pi}{a}\right)^2 + \left(\frac{\pi}{b}\right)^2 + \left(\frac{\pi}{c}\right)^2$	$\phi_m\cos\left(\frac{\pi x}{a}\right)\cos\left(\frac{\pi y}{b}\right)\cos\left(\frac{\pi z}{c}\right)$	$\left(\frac{2}{\pi}\right)^3$	$\frac{161}{B^3}$
Infinite cylinder, radius R	$\left(\frac{2.405}{R}\right)^2$	$\phi_m J_0\left(\frac{2.405r}{R}\right)$	0.432	—
infinite slab, thickness a	$\left(\frac{\pi}{a}\right)^2$	$\phi_m\cos\left(\frac{\pi x}{a}\right)$	$\frac{\pi}{2}$	—

* All dimensions include extrapolation distances.

> Case B.   Homogeneous core with reflector
>
> One group diffusion method
>
> Steady state

Starting with the wave equations for the core region c and the reflector region r, one has

$$\begin{cases} D_c \nabla^2 \phi_c - \Sigma_{ac} \phi_c + S_c = 0 & \text{(core)} \\ D_r \nabla^2 \phi_r - \Sigma_{ar} \phi_r = 0 & \text{(reflector)} \end{cases}$$

or,

$$\begin{cases} \nabla^2 \phi_c + B_c^2 \phi_c = 0 & \text{(core)} \qquad (6\text{-}5) \\ \nabla^2 \phi_r - \kappa_r^2 \phi_r = 0 & \text{(reflector)} \qquad (6\text{-}6) \end{cases}$$

where B_c^2 is defined by equation 5-13, $\kappa_r^2 = \Sigma_{ar}/D_r = 1/L_r^2$, and there is no source term in the reflector region, since no thermal neutron is produced there, according to the one group theory. The boundary conditions are:

a)   The flux must be finite everywhere and be symmetric.

b)   The flux ϕ must be contineous across the core-reflector boundary.

c)   The current density j must be contineous across the core-reflector boundary.

d)   The flux must vanish at the outside surface (extrapolated) of the reflector.

Consider, for example, a radially reflected right cylindrical reactor with core radius R_c and reflector thickness $(R_r - R_c)$. For this case, the z-dependent part of the flux is just like that of case A and the problem is reduced to an one-dimentional case.

Equation 6-5 is now written as

$$\nabla^2 \phi_c(r) + B_R^2 \phi_c(r) = 0 \qquad (6\text{-}7)$$

where r is the distance from the axis (not from the origin) and

$$B_R^2 = B_c^2 - B_z^2 = B_c^2 - \left(\frac{\pi}{H_0}\right)^2$$

Thus, (see solved problem 6-1)

$$\phi_c(r) = A_c J_0(B_R r) \qquad (6\text{-}8)$$

where A_c is a constant and boundary condition (a) has been utilized.

Similiarly, for

$$\kappa_R^2 = \kappa_r^2 + B_z^2 = \kappa_r^2 + \left(\frac{\pi}{H_0}\right)^2$$

one has

$$\nabla^2 \phi_r(r) - \kappa_R^2 \phi_r(r) = 0 \qquad (6\text{-}9)$$

and the solution of equation 6-9 is (see solved problem 6-1)

$$\phi_r(r) = A_r I_0(\kappa_R r) + C_r K_0(\kappa_R r) \tag{6-10}$$

Boundary condition (d), $\phi_r(R_r) = 0$, provides

$$C_r = -A_r \frac{I_0(\kappa_R R_r)}{K_0(\kappa_R R_r)}$$

as a mean to eliminate C_r from equation 6-10.

Boundary condition (b), $\phi_c(R_c) = \phi_r(R_c)$, and boundary condition (c),

$$-D_c \frac{\partial \phi_c(r)}{\partial r}\bigg|_{r=R_c} = -D_r \frac{\partial \phi_r(r)}{\partial r}\bigg|_{r=R_c}, \qquad \text{provide two more equations to}$$

eliminate the unknown proportional constants A_c and A_r, and one has, finally,

$$\frac{D_r \kappa_R J_0(B_R R_c)}{D_c B_R J_1(B_R R_c)} = \frac{I_0(\kappa_R R_r) K_0(\kappa_R R_c) - I_0(\kappa_R R_c) K_0(\kappa_R R_r)}{I_1(\kappa_R R_c) K_0(\kappa_R R_r) + I_0(\kappa_R R_r) K_1(\kappa_R R_c)} \tag{6-11}$$

where J_1 and I_1 are given by

$$J_1(x) = -\frac{d}{dx} J_0(x)$$

$$I_1(x) = \frac{d}{dx} I_0(x)$$

Equation 6-11 is the criticality condition for a radially reflected cylindrical reactor, from which one can solve for any one unknown such as R_r, D_c, or B_R. (It is no longer true that $B^2 = (2.405/R_c)^2$, as in the case of bare core.) Figure 6.2 is a plot of the flux distributions in the radial direction, based on one group theory. Similar treatments of the one-group multi-region problems lead to the results of Table 6.2.

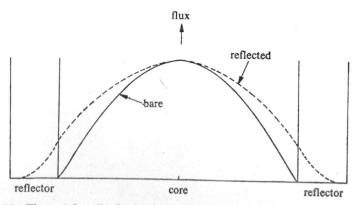

Figure 6.2 Thermal flux distribution in bare and reflected homogeneous reactors, based on one-group theory.

Reactor	Criticality condition	Flux
Rectangular block, core dimensions a, b, c. Reflector thickness T on each of the two opposite sides of the z-direction only.	$D_c B_z \tan\left(\dfrac{B_z c}{2}\right) = D_r \kappa_z \coth(\kappa_z T)$ where $B_z^2 = B_c^2 - \left(\dfrac{\pi}{a}\right)^2 - \left(\dfrac{\pi}{b}\right)^2$ $\kappa_z^2 = \kappa_r^2 + \left(\dfrac{\pi}{a}\right)^2 + \left(\dfrac{\pi}{b}\right)^2$	$\phi_c = \phi_m \cos\left(\dfrac{\pi x}{a}\right)\cos\left(\dfrac{\pi y}{b}\right)\cos(B_z z)$ $\phi_r = \phi_m \dfrac{\cos(B_z c/2)}{\sinh(\kappa_z T)}\cos\left(\dfrac{\pi x}{a}\right)\cos\left(\dfrac{\pi y}{b}\right)\sinh\left[\kappa_z\left(\dfrac{c}{2}-z\right)\right]$
Spherical core radius R_c, reflector thickness $(R_r - R_c)$.	$D_c[B_c R_c \cot(B_c R_c) - 1] =$ $- D_r\left[\dfrac{R_c}{L_r}\coth\left(\dfrac{R_r - R_c}{L_r}\right) + 1\right]$	$\phi_c = \phi_m \dfrac{\sin(B_c r)}{r}$ $\phi_r = \phi_m \dfrac{\sin(B_c R_c)}{\sinh\kappa_r(R_r - R_c)}\dfrac{\sinh\kappa_r(R_r - r)}{r}$
Cylindrical core radius R_c, height H_0, radial reflector thickness $(R_r - R_c)$, and origin at the middle of the axis.	equation 6-11	$\phi_c = \phi_m J_0(B_{RR})\cos\left(\dfrac{\pi z}{H_0}\right)$ $\phi_r = \phi_m \dfrac{J_0(B_R R_c)}{M_0(R_c)}M_0(r)\cos\left(\dfrac{\pi z}{H_0}\right)$ where $B_R^2 = B_c^2 - (\pi/H_0)^2$ $M_0(r) = I_0(\kappa_r R_r)K_0(\kappa_r r) - K_0(\kappa_r R_r)I_0(\kappa_r r)$
Cylindrical core radius R_c, height H_0, end reflectors thickness T at each of the two ends only, and origin at the middle of the axis.	$D_c B_z \tan\left(\dfrac{B_z H_0}{2}\right) = D_r \kappa_z \coth(\kappa_z T)$ where $B_z^2 = B_c^2 - \left(\dfrac{2.405}{R_c}\right)^2$ $\kappa_z^2 = \kappa_r^2 + \left(\dfrac{2.405}{R_c}\right)^2$	$\phi_c = \phi_m J_0(2.405 r/R_c)\cos(B_z z)$ $\phi_r = \phi_m \dfrac{\cos(B_z H_0/2)}{\sinh(\kappa_z T)}J_0\left(\dfrac{2.405 r}{R_c}\right)\sinh\left[\kappa_z\left(\dfrac{H_0}{2}+T-z\right)\right]$

Subscripts: c = core, r = reflector, m = origin.
Reflector thicknesses include extrapolation distances.

> Case C.   Homogeneous core with reflector
>
> Two-group diffusion method
>
> Steady state

In this model, the neutrons are assumed to be in two energy groups, the fast and the thermal. The following subscripts will be used:

f = fast group

t = thermal group

c = core

r = reflector

a = thermal absorption

b = fast removal (due to absorption and scattering to lower energy).

The processes in the core and reflector are summarized in Figure 6.3.

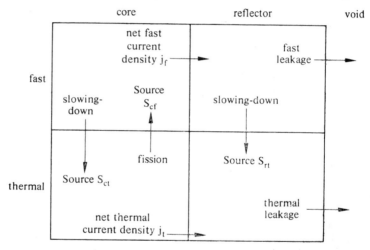

Figure 6.3   Neutron balance processes in two-group two-region reactor.

The neutron balance equations for the two groups and two regions are:

$$\left\{\begin{array}{ll} D_{cf}\nabla^2\phi_{cf} - \Sigma_{cb}\phi_{cf} + \phi_{ct}\Sigma_{ca}f\eta\epsilon = 0 & \text{(6-12)} \\[4pt] D_{ct}\nabla^2\phi_{ct} - \Sigma_{ca}\phi_{ct} + \phi_{cf}\Sigma_{cb}p = 0 & \text{(6-13)} \\[4pt] D_{rf}\nabla^2\phi_{rf} - \Sigma_{rb}\phi_{rf} + 0 = 0 & \text{(6-14)} \\[4pt] D_{rt}\nabla^2\phi_{rt} - \Sigma_{ra}\phi_{rt} + \phi_{rf}\Sigma_{rb} = 0 & \text{(6-15)} \end{array}\right.$$

in which $\Sigma_{cb} = D_{cf}/\tau_c$ and $\Sigma_{rb} = D_{rf}/\tau_r$, similar to $\Sigma_{ca} = D_{ct}/L_c^2$.

For the core region, one can write

$$\nabla^2\phi_{cf} + B^2\phi_{cf} = 0 \tag{6-16}$$

$$\nabla^2\phi_{ct} + B^2\phi_{ct} = 0 \tag{6-17}$$

provided one sets

$$(1 + B^2\tau_c)\phi_{cf} - (\Sigma_{ca}\tau_c f\eta\epsilon/D_{cf})\phi_{ct} = 0 \tag{6-18}$$

$$\left(-\frac{pD_{cf}}{\Sigma_{ca}\tau_c}\right)\phi_{cf} + (1 + B^2L_c^2)\phi_{ct} = 0 \tag{6-19}$$

so that equations 6-12 and 6-13 are not violated.

For equations 6-18 and 6-19 to be true for all ϕ_{cf} and ϕ_{ct}, the determinant of their coefficients must be zero. This leads to

$$(1 + B^2\tau_c)(1 + B^2L_c^2) = pf\eta\epsilon = k_\infty \tag{6-20}$$

Equation 6-20 is not the criticality condition for a reflected reactor. It does not involve the reflector and is merely a restriction on the value of B^2 such that equations 6-12, 6-13, 6-16, and 6-17 are true simultaneously.

The solutions of equation 6-20 for B^2 have one positive value and one negative value and will be designated as B_1^2 and $-B_2^2$. Their values depend on reactor materials rather than geometry. Thus,

$$\phi_{cf} = A_1P_c + C_1Q_c \tag{6-21}$$

$$\phi_{ct} = A_2P_c + C_2Q_c \tag{6-22}$$

where P_c and Q_c are functions of positions that satisfy the equations

$$\nabla^2P_c + B_1^2P_c = 0 \tag{6-23}$$

$$\nabla^2Q_c + B_2^2Q_c = 0 \tag{6-24}$$

and A_1, A_2, C_1, and C_2 are constants.

Substituting equations 6-21 and 6-22 into equation 6-12 and applying equations 6-23 and 6-24, one gets

$$(-D_{cf}A_1B_1^2 - A_1\Sigma_{cb} + A_2\Sigma_{ca}f\eta\epsilon)P_c$$
$$+ (D_{cf}C_1B_2^2 - C_1\Sigma_{cb} + C_2\Sigma_{ca}f\eta\epsilon)Q_c = 0$$

In general, P_c and Q_c are independent functions. The coefficients in the equation above must therefore be set to zero. This gives

$$A_2 = \frac{D_{cf}B_1^2 + \Sigma_{cb}}{\Sigma_{ca}f\eta\epsilon}A_1 = S_1A_1 \tag{6-25}$$

$$C_2 = \frac{-D_{cf}B_2^2 + \Sigma_{cb}}{\Sigma_{ca}f\eta\epsilon}C_1 = S_2C_1 \tag{6-26}$$

G

Thus, only two of the four constants, namely A_1 and C_1, are to be determined.

The reflector equations 6-14 and 6-15 are easier to solve, since they are not coupled together as equations 6-12 and 6-13 are through the source terms. By solving the *homogeneous* equation 6-14 directly and eliminating one of the two resulting proportional constants by the boundary condition that the *fast* flux must vanish at the outside extrapolated surface of the reflector, one gets

$$\phi_{rf} = A_3 U_r \qquad (6\text{-}27)$$

where A_3 is a constant and U_r is a function of position that satisfies

$$D_{rf}\nabla^2 U_r - \Sigma_{rb}U_r = 0 \qquad (6\text{-}28)$$

Likewise, the solution of the homogeneous part of the *inhomogeneous* equation 6-15 can be written as $A_4 U_r$, similarly making use of the boundary condition that the *thermal* flux must vanish at the same outside extrapolated boundary of the reflector as the fast flux. Now, if a particular solution for equation 6-15 is written as $C_3 V_r$, *i.e.*, if

$$D_{rt}\nabla^2 V_r - \Sigma_{ra}V_r + \Sigma_{rb}A_3 U_r = 0 \qquad (6\text{-}29)$$

and

$$\phi_{rt} = A_4 U_r + C_3 V_r \qquad (6\text{-}30)$$

then, by substituting equation 6-30 into equation 6-15 and applying equations 6-28 and 6-29, one gets

$$A_4 = \frac{\Sigma_{rb}}{(D_{rt}/D_{rf})\Sigma_{rb} - \Sigma_{ra}} A_3 = S_3 A_3 \qquad (6\text{-}31)$$

The fast and thermal fluxes in the core and reflector are now theoretically determined and the only remaining job is to determine the four constants, A_1, C_1, A_3, and C_3. The four boundary conditions that the fast and thermal fluxes and the fast and thermal current densities must be continuous across the core-reflector interface are used. This gives:

$$\begin{cases} (P)A_1 + (Q)C_1 - (U)A_3 & = 0 \quad (6\text{-}32) \\ (S_1 P)A_1 + (S_2 Q)C_1 - (S_3 U)A_3 - (V)C_3 & = 0 \quad (6\text{-}33) \\ (D_{cf}P')A_1 + (D_{cf}Q')C_1 - (D_{rf}U')A_3 & = 0 \quad (6\text{-}34) \\ (S_1 D_{ct}P')A_1 + (S_2 D_{ct}Q')C_1 - (S_3 D_{rt}U')A_3 - (D_{rt}V')C_3 = 0 \quad (6\text{-}35) \end{cases}$$

where

A_1, C_1, A_3, and C_3 are unknowns;

P, Q, U, and V are values of P_c, Q_c, U_r, and V_r, respectively, evaluated at the core-reflector boundary;

P', Q', U', and V' are derivatives of P, Q, U, and V, respectively, with respect to the normal coordinate at the boundary and evaluated at the boundary.

The set of four *homogeneous* equations are consistent only if the determinant Δ of the coefficients is zero. This gives, finally, the criticality condition for a reflected reactor, based on two-group model:

$$\Delta = 0 \quad \text{(criticality condition)} \qquad (6\text{-}36)$$

Since the four equations 6-32 through 6-35 are homogeneous, it is not possible to solve for all four unknowns. Rather, only three may be explicitly solved for. The remaining proportional constant will, of course, depend on the power level of the reactor.

Figure 6.4 is a plot of the fast and thermal fluxes. The thermal neutron "hump" in the reflector is due to the fact that in the reflector slow neutrons are produced by the slowing-down of fast ones, but they are absorbed less strongly in the reflector than in the core.

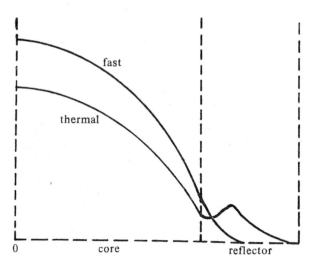

Figure 6.4   Fast and thermal fluxes in a reflected homogeneous reactor, based on two-group theory.

The flux functions P_c, Q_c, U_r, and V_r for three different geometries are given in Table 6.3. In case of cylindrical geometry, for example, if the height is not infinite and the ends are not reflected, it is, as before, necessary to use

TABLE 6.3   Two-group flux functions.

	Spherical core radius R_c, reflector thickness T	Cylindrical core radius R_c, radial reflector thickness T	Infinite slab with core thickness H, reflector thickness T on each side
P_c	$\dfrac{\sin(B_1 r)}{r}$	$J_0(B_1 r)$	$\cos(B_1 x)$
Q_c	$\dfrac{\sinh(B_2 r)}{r}$	$I_0(B_2 r)$	$\cosh(B_2 x)$
U_r	$\dfrac{\sinh[\kappa_{rf}(R_c + T - r)]}{r}$	$K_0(\kappa_{rf}r) - \dfrac{K_0[\kappa_{rf}(R_c + T)]}{I_0[\kappa_{rf}(R_c + T)]} I_0(\kappa_{rf}r)$	$\sinh\left[\kappa_{rf}\left(\dfrac{H}{2} + T - x\right)\right]$
V_r	$\dfrac{\sinh[\kappa_{rt}(R_c + T - r)]}{r}$	$K_0(\kappa_{rt}r) - \dfrac{K_0[\kappa_{rt}(R_c + T)]}{I_0[\kappa_{rt}(R_c + T)]} I_0(\kappa_{rt}r)$	$\sinh\left[\kappa_{rt}\left(\dfrac{H}{2} + T - x\right)\right]$

Reflector thickness include extrapolated distance.

the modified bucklings $B_{1R}^2 = B_1{}^2 - (\pi/H)^2$ and $B_{2R}^2 = B_2{}^2 + (\pi/H)^2$ instead of $B_1{}^2$ and $B_2{}^2$ in Table 6.3.

It should be noted that although the analysis given in this section is developed theoretically for all geometries (since no geometrical restrictions were imposed) only certain geometries such as those listed in Table 6.2 were analytically possible. For the most important case of a completely reflected cylinder (*i.e.*, finite core completely covered by reflector) as in the case of large power reactor, the method of separation of variables developed here cannot be used. Nevertheless, by using iterative type of calculations and treating the radially reflected and end reflected cases alternatively in conjunction with the "equivalent-bare-reactor" concept, rough reactor calculations can be made by disregarding "corner effects". This is, however, beyond the scope of this text.

Case D.   Heterogeneous cell
One-group diffusion method
Steady state

Having seen the general features of the flux distributions in a reactor, attention is now focused on the 'fine structure' and the characteristics of a heterogeneous cell. The advantages and disadvantages of a heterogeneous system are:

1) Increase of resonance escape probability p—By lumping the fuel, moderator, and other materials in a heterogeneous manner, the resonance escape probability usually increases. This is because neutrons suffer little energy loss in collision with the fuel. Only those neutrons with energies close to the resonance peaks will get absorbed. The rest will pass through the fuel region and into the moderator region where the neutrons will be slowed down to energies below the resonance peaks. Those neutrons that were absorbed were absorbed in the fuel near the fuel-moderator boundary. The majority of the fuel nuclei in the lump are therefore exposed to a depressed flux. (The average thermal neutron flux in the moderator is larger than that in the fuel where absorption is larger) and give rise to a larger p.

2) Increase of fast fission factor ϵ—The fast fission neutrons produced in the fuel suffer little energy loss in the fuel and therefore have a better chance of causing fast fission.

3) Provide a convenient design of coolant channel for heat removal.

4) Decrease in thermal utilization f—Since most slowing-down occurs in the moderator, the average thermal neutron flux is higher there than in the

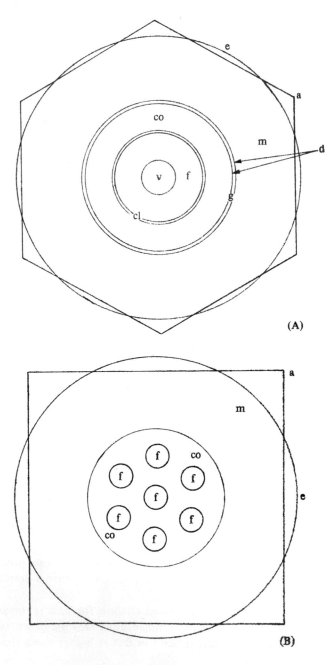

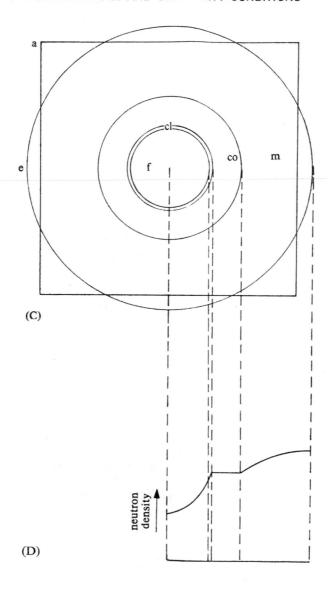

(C)

(D)

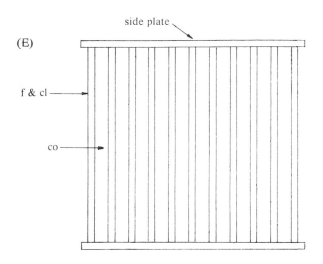

Figure 6.5   Cross-sectional views of various heterogeneous cells.
a = actual cell boundary   cl = cladding   co = coolant   d = duct metal
e = equivalent cell boundary   f = fuel   g = gas insulation   m = moderator
v = void (for fission gas)

fuel. As a result, f is lower than that of a homogeneous system. However, the decrease in f is usually more than counterbalanced by the increase in p.

In the study of heterogeneous reactors, one is interested in:
1)   Analysing the cell and find the flux at every location.
2)   Homogenize the core and find effective values for f, p, ϵ, τ, etc. so that a criticality condition for the heterogeneous reactor may be found.

Heterogeneous cell may take many different geometries. Figure 6.5 shows a few of them. In general, the following assumptions are made as a base for rough analysis:
1)   One-group diffusion theory is applicable in both the fuel and the moderator.
2)   Cell is representative of system.
3)   Flux does not vary in axial direction (good approximation for long cell).
4)   Slowing-down density q is constant throughout the moderator and zero in the fuel (*i.e.*, there is no thermal neutron source in the fuel).
5)   The flux is flat at the cell boundary (*i.e.*, net neutron current density is zero there.)

One has, therefore, a one-group multi-region problem. The flux inside a cell can be solved for in conjunction with the appropriate geometry and boundary conditions. A plot of the neutron density for Figure 6.5C is given in Figure 6.5D. With the flux known, various reactor constants can be found:

1) Thermal Utilization f

The thermal utilization, defined as the ratio of absorption in fuel and total absorption, must be weighted by the appropriate fluxes in different regions of a heterogeneous cell and is written as

$$f = \frac{\int_F \phi_F \Sigma_{aF}\, dV}{\int_F \phi_F \Sigma_{aF}\, dV + \int_m \phi_m \Sigma_{am}\, dV} \tag{6.37}$$

where the subscripts

   a = thermal absorption
   F = fuel (U-235 and U-238, for example)
   m = materials (such as moderator, coolant, structural materials, etc.) other than fuel.

2) Resonance-escape Probability p

Recalling equation 5-21 for p of a homogeneous reactor,

$$p = \exp\left[-\int_E^{E_0} \frac{\Sigma_a}{\xi\Sigma_t} \frac{dE}{E} \right]$$

$$= \exp\left[-\int_{u_0}^u \frac{\Sigma_a}{\xi\Sigma_t}\, du \right] = \exp\left[-\frac{\Sigma_a}{\xi\overline{\Sigma}_t} \right] \tag{5-21}$$

When diffusion theory is used, the medium is assumed to have weak absorption in comparison with scattering. Thus, for a homogeneous reactor,

$$p = \exp\left[-\frac{\Sigma_a}{\xi\overline{\Sigma}_s} \right] \qquad \text{(homogeneous)}$$

Again, for heterogeneous reactor, the flux weighting must be applied and, since absorption is mainly in the fuel and scattering mainly outside of the fuel, one has, for the heterogeneous case,

$$p = \exp\left[-\frac{\int_F \phi_F \overline{\Sigma}_{ar}\, dV}{\xi \int_m \phi_m \overline{\Sigma}_{sr}\, dV} \right] \qquad \text{(heterogeneous)} \tag{6-38}$$

where the 'bar' and the subscript r emphasize that the cross-sections are averaged over the resonance region, not the thermal region.

3) Age τ and Diffusion Length L

There are many factors that make τ and L of a heterogeneous cell different from that of a moderator. For example, it is easier for both fast and thermal neutrons to move parallel to the fuel and the void, if any, than to cross it perpendicularly. This streaming or channeling effect, together with other factors such as different elastic and inelastic scattering properties of the fuel and the moderator make it impossible to consider the age and the diffusion length of a heterogeneous cell in detail at this level. However, for rough estimates, one may use

$$L^2 \approx L_m{}^2(1 - f)\, \frac{V_m + V_F(\bar{\phi}_F/\bar{\phi}_m)}{V_m} \tag{6-39}$$

$$\tau \approx \tau_m \left(\frac{V_m + V_F}{V_F}\right)^2 \tag{6-40}$$

where f is the thermal utilization factor and the subscripts are the same as that of equation 6-37.

4) Fast Fission Factor ϵ

The fast fission factor of a heterogeneous cell is related to the probability P (to be derived later) that a fast neutron born in the fuel will make a collision with a fuel nucleus before escaping into the moderator.

Consider a primary neutron that has energy above the threshold of fission. Then, in one generation,

Collision in the fuel $= P$
Fission neutrons produced $= v\sigma_f P/\sigma$
Elastic collisions in the fuel $= \sigma_e P/\sigma$
Neutrons escaping from the fuel without a collision $= 1 - P$
Neutrons slowed down below threshold by inelastic collision $= \sigma_i P/\sigma$

where all the cross-sections and v are for fast neutrons. In other words, $x = (1 - P + \sigma_i P/\sigma)$ neutrons slow past the threshold and $[(v\sigma_f/\sigma) + (\sigma_e/\sigma)]P = yP$ neutrons will be available for the second and higher generations. These neutrons, on the average, will have a different probability P' of collision with the fuel since they are distributed uniformly in the fuel and are not like the depressed thermal flux which produced the first generation primary fission neutrons.

Since ϵ is the total number of neutrons that slow past the threshold per primary fission neutron, one can get ϵ by summing up all the generations.

Evidently,

$$\epsilon = x + yPx + (yP')(yPx) + (yP')^2(yPx) + (yP')^3(yPx) + \ldots.$$

$$= x + (yPx)[(yP') + (yP')^2 + (yP')^3 + \ldots.]$$

$$\epsilon = x + (yPx)\left[\frac{yP'}{1 - yP'}\right] \tag{6-41}$$

It remains to find P and P'. The following notations will be used (see Figure 6.6):

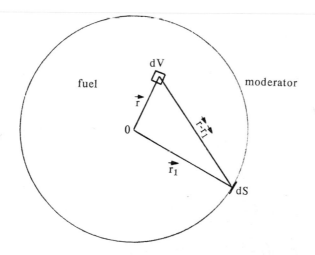

Figure 6.6   Geometry for calculating fast fission factor

Σ_f = fission cross-section for thermal neutrons.

$\phi_F(r)$ = thermal neutron flux at position $\mathbf{r}$ in the fuel so that

$\Sigma_f v \phi_F(r)\, dV$ = fission neutrons production rate within differential volume element dV located at $\mathbf{r}$.

Σ = total cross-section for fast neutrons.

dS = differential surface area of the fuel.

The number of neutrons that is created in dV and escape through dS *without* a collision is

$$\Sigma_f v \phi_F(r)\, dV \frac{\exp(-\Sigma|\mathbf{r}_1 - \mathbf{r}|)}{4\pi |\mathbf{r}_1 - \mathbf{r}|^2} \cos\theta\, dS$$

where $\exp(-\Sigma|\mathbf{r}_1 - \mathbf{r}|)$ is an attenuation factor due to Σ, $1/(4\pi|\mathbf{r}_1 - \mathbf{r}|^2)$

is an attenuation factor due to the distance involved, and the factor $\cos \theta$ makes the area $\cos \theta \, dS$ an effective area perpendicular to the direction $(\mathbf{r}_1 - \mathbf{r})$.

The probability P that a primary neutron *will* have a collision inside the fuel is thus

$$P = 1 - \frac{\displaystyle \int_S \int_V \Sigma_f v \phi_F(r) \frac{\exp(-\Sigma |\mathbf{r}_1 - \mathbf{r}|)}{4\pi |\mathbf{r}_1 - \mathbf{r}|^2} \cos \theta \, dV \, dS}{\displaystyle \int_V \phi_F(r) \Sigma_f v \, dV} \tag{6-42}$$

where the second term on the right is the escape probability of neutrons *without* a collision. The denominator is the total primary neutron production rate. For P', one simply sets $\phi_F(r) = \bar{\phi}_F = $ constant. Figure 6.7 is a plot of P' for a cylindrical fuel rod.

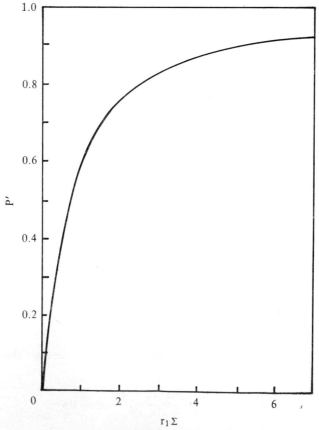

Figure 6.7 Collision probability P' of neutrons in a cylindrical rod.

5) *Criticality Condition*

Generally speaking, the criticality conditions derived for the homogeneous cases are applicable to heterogeneous systems, provided that the heterogeneous systems are homogenized and the reactor constants above are used. The task of adjusting all the constants is, however, more or less an art of the experienced engineers and, in this respect, one must feel humble when one thinks of the Manhattan Project and the masters involved.

SOLVED PROBLEMS

6-1 Explain how equations 6-4, 6-8, and 6-10 are obtained.

Answer: Equation 6-1 is called a Bessel equation of zero order when a^2 is positive. The solutions of the equation are represented by two functions, $J_0(ar)$ and $Y_0(ar)$, called the zero order Bessel functions of the first and second kinds, respectively. The two functions are plotted in Figure 6.8.

When a^2 is negative, equation 6-1 is called a modified Bessell equation of zero order. The solutions are $I_0(ar)$ and $K_0(ar)$, called the zero order modified Bessel functions of the first and second kinds, respectively. These two functions are also plotted in Figure 6.8.

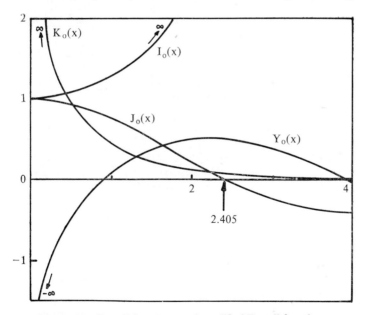

Figure 6.8 Bessell functions and modified Bessell functions.

For the case of a bare reactor core, the requirement that the flux must be finite at $r = 0$ eliminates K_0 and Y_0 as solutions. Since $I_0(ar)$ increases rapidly with r and it is not possible to make any linear combination of $I_0(ar)$ and $J_0(ar)$ to be positive at the origin and vanish at the boundary, $I_0(a_2)$ is also eliminated. This leaves $J_0(ar)$ and a^2 is positive.

Consider now equation 6-2. If β^2 is negative, the solutions would be exponential in form and it is not possible to get a linear combination of $Ae^z + Be^{-z}$ to be symmetric with respect to the $z = 0$ plane, remain finite inside the core, and vanish at the boundary. Thus, β^2 cannot be negative. For positive β^2, the solutions of equation 6-2 are sin (βz) and cos (βz). Since sin (βz) is not symmetric with respect to the $z = 0$ plane, it is eliminated, leaving only cos (βz).

The condition that the flux must be zero at the bare core boundary requires that $\beta = \pi/H_0$ and $a = 2.405/R_0$ and one has equation 6-4 for a bare core.

In the case of a reflected reactor, the flux in the core must have the same form as the bare core, since the former case must be valid even for zero reflector thickness. Thus, one has equation 6.8.

Equation 6-10 is obtained because $\kappa_R{}^2 = \Sigma_{ar}/D_r + (\pi/H_0)^2$ is positive. There is no reason to eliminate either I_0 or K_0 since the reflector does not extend back to the origin and it is permissible for K_0 to "blow-up" there, as far as the reflector flux is concerned. As is seen in the text, it is possible to adjust the coefficients A_r and C_r such that the reflector flux vanishes at the outside extrapolated boundary.

6-2   For a bare homogeneous cylindrical core, what is the ratio of height to diameter, if the core has:

a)   minimum critical volume for a given $B_g{}^2$?

b)   minimum surface area for a given volume?

c)   maximum $B_g{}^2$ for a given volume?

Answer:
Note: All dimensions below are extrapolated dimensions.
a)   In the discussion of minimum critical volume, one has

$$H = \frac{\pi\sqrt{3}}{B_g} \quad \text{and} \quad R = \frac{2.405}{B_g}\sqrt{\frac{3}{2}}$$

Thus,

$$\frac{H}{2R} = 0.9235$$

b)

$$S = 2\pi RH + 2\pi R^2 = 2\pi R\left(\frac{V}{\pi R^2}\right) + 2\pi R^2$$

$dS/dR = 0$ gives $H/(2R) = 1$

c) $d(B_g{}^2)/dR = 0$ leads to $H/2R = 0.9235$. This is the same result as in part (a) and is expected since the minimum critical volume V_{min} is related to $B_g{}^2$ by $V_{min} \propto 1/B_g{}^3$. When one is an extremum, the other is too.

6-3   For a thermal reactor, what is the difference between one-group diffusion theory and age-diffusion theory?
Answer: In the one-group diffusion theory, one assumes that all fission neutrons are thermalized on the spot as soon as they are born. In other words, $\tau = 0$ and there is no

fast leakage. The neutrons are assumed to have one effective energy near the thermal energy and diffusion theory is assumed. For a bare homogeneous core, the criticality condition is

$$\frac{k_\infty}{1 + L^2 B^2} = 1$$

In the one group age diffusion theory, immediate thermalization is not assumed. The criticality condition is

$$\frac{k_\infty e^{-B^2 \tau}}{1 + L^2 B^2} = 1$$

It is interesting to note that

$$\ln (1 + L^2 B^2) = (\ln k) - B^2 \tau$$

For $L^2 B^2 \ll 1$, $\ln (1 + L^2 B^2) \approx L^2 B^2$, and for $k_\infty \approx 1$, $\ln k_\infty \approx k_\infty - 1$. Thus,

$$L^2 B^2 \approx (k_\infty - 1) - B^2$$

or,

$$\frac{k_\infty}{1 + (L^2 + \tau) B^2} \approx 1$$

This is similar to the one-group theory above. When $(L^2 + \tau)$ is used instead of L^2, it is often referred to as the "modified one-group theory". The term $(L^2 + \tau)$ has a dimension of area and is often referred to as migration area M^2.

SUPPLEMENTARY PROBLEMS

6-a Find the flux distribution of a spherical homogeneous bare reactor with a concentric hole in the middle, using one-group diffusion model. Note that j is zero at the inner boundary.

6-b Given the following constants for a spherical homogeneous reactor:
$D_{cf} = D_{ct} = D_{rf} = D_{rt} = 1$ cm
$\Sigma_{cf} = \Sigma_{rf} = 0.01$ cm^{-1} ⎫
$\Sigma_{ct} = 0.05$ cm^{-1} ⎬ absorption cross-sections
$\Sigma_{rt} = 0.005$ cm^{-1} ⎭
$\Sigma_f = 0.023$ cm^{-1} (thermal fission cross-section in the core)
$p_c = p_r = 0.9$
$\epsilon = 1.02$
$T = 20$ cm (extrapolated)
and given the following models:
  1. Bare core, 1 group
  2. Reflected core, 1 group
  3. Bare core, modified one group

    4. Reflected core, modified one group
    5. Bare core, two group
    6. Reflected core, two group
Find the critical radii using
    models 1, 3, and 4
    or   2 and 5
    or   6

Note: Models 1 and 2 should give R between 60 cm and 80 cm. The other four models should give R between 180 cm and 200 cm.

Reference: "Reactor Analysis" by Meghreblian and Holmes, P. 467.

Chapter 7

REACTOR KINETICS AND CONTROL

IN THIS chapter, the kinetic behavior of a reactor is presented, followed by a discussion of control rod theory and the factors that affect the stability and control of a reactor.

A. Reactor Kinetics

1. Kinetic Equations

Assuming that the space and time-dependent flux distribution is separable, *i.e.*, if the shape of the neutron flux remains the same as is the case with uniform disturbance and only the amplitude changes with time, equation 5-11 can be written as

$$\frac{dn}{dt} = D\nabla^2\phi - \Sigma_a\phi + S$$

If the reactor is near critical, $\nabla^2\phi = -B^2\phi$, and one has

$$\frac{dn}{dt} = -(DB^2 + \Sigma_a)\phi + S = -(1 + B^2L^2)\Sigma_a\phi + S = -\frac{v\Sigma_a}{L_t}n + S$$

Since the thermal diffusion time

$$t_t = \frac{\text{number of thermal neutrons present}}{\text{rate of thermalization}} = \frac{n}{k_\infty\phi\Sigma_aL_f} = \frac{L_t}{v\Sigma_a}$$

one has

$$\frac{dn}{dt} = -\frac{n}{t_t} + S$$

where $n = n(t)$ and the source S is due to prompt neutrons S_p and delayed neutrons S_d. If β represents the fraction of neutrons that are delayed (Table 3.4) and if the concentration of available delayed neutrons that reach thermal energies is designated as C_1 for the appropriate group i with decay constant λ_i (Table 3.3), then

$$S = S_p + S_d$$
$$S_p = k_\infty\phi\Sigma_a(1 - \beta)L_f$$
$$S_d = \sum_i^6 \lambda_iC_i$$

H

By defining the excess multiplication factor

$$\delta k_e = k_e - 1$$

and by defining the prompt excess multiplication factor

$$\delta k_{ep} = k_{ep} - 1 = k_e(1 - \beta) - 1 = \delta k_e - \beta k_e \neq \delta k_e(1 - \beta)$$

as the excess multiplication factor due to prompt neutrons, one has, by simple substitutions,

$$\frac{dn}{dt} = \frac{\delta k_{ep}}{t_t} n + \sum_i^6 \lambda_i C_i$$

A more rigorous analysis, taking into account the fact that the source S must be evaluated at a time earlier than t by the slowing down time t_s (see Table 5.1 and note that $t_s \ll t_t$) leads to

$$\frac{dn}{dt_t} = \frac{\delta k_{ep}}{l} n + \sum_i^6 \lambda_i C_i \qquad (7\text{-}1)$$

where $l = t_s + t_t$ is the total life-time of a neutron.*

The concentration C_i is coupled to n by

$$\frac{dC_i}{dt} = -\lambda_i C_i + \frac{k_e \beta_i}{l} n \qquad (7\text{-}2)$$

since the production of delayed neutrons is governed by the change in concentration of the parent nuclei.

Equations 7-1 and 7-2 are the reactor kinetic equations.

2. *Inhour Equation and Reactivity*

To solve equations 7-1 and 7-2 for a constant δk_e, one tries by assuming that both $n(t)$ and $C_i(t)$ are proportional to $e^{+t/T}$ where T is a constant and is called the period. (It is not to be confused with the neutron generation time.) The assumption leads to

$$C_i = \frac{k_e \beta_i n}{l(1/T + \lambda_i)} \qquad (7\text{-}3)$$

and

$$\frac{\delta k_e}{k_e} = \frac{l}{k_e T} + \sum_i^6 \frac{\beta_i}{1 + \lambda_i T} \qquad (7\text{-}4)$$

where $\lambda = 1/\tau_i$

---

* However, l is not the generation time, since some of the neutrons are delayed. The generation time l^* is given by $l^* = l + \beta\bar{\tau}$ where $\bar{\tau}$ is defined in problem 3-6 and should not be confused with the Fermi age.

Equation 7-4 is called the "inhour equation"* and $\rho = \delta k_e / k_e$ is called the reactivity. The reactivity may have two different units† (just like a length may have foot or centimeter as unit):

a) *Dollar*—A dollar of reactivity is when $\rho = \beta$. A cent of reactivity is when $\rho = 0.01\beta$

b) *Inhour*—An inhour of reactivity is when ρ is such that $T = 1$ hour. This unit is used infrequently.

Figure 7.1 gives the qualitative appearance of ρ vs. T. It is noted that negative values of T lead to decreasing neutron density as time goes on, since $n \propto e^{t/T}$. A negative reactivity would therefore always lead to a reactor

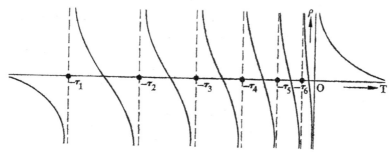

Figure 7.1 Qualitative plot of reactivity ρ vs. period T of the inhour equation.

shutdown. For positive ρ and after a long time (in comparison with τ_1), only the period $T > 0$ is important. This period is often referred to as the "stable period". In fact, if

a) $\rho \gg \beta$, then $T \to 0$ and $\lambda_i T \ll 1$. Equation 7-4 becomes

$$\rho = \frac{l}{k_e T} + \beta$$

or,

$$T = \frac{l}{\delta k_{ep}}$$

and

$$n(t) = n_0 \exp\left(+\frac{\delta k_{ep}}{l}t\right)$$

---

* If the fact that delayed neutrons are produced at a lower energy (0.5 Mev) than fission neutrons (2 Mev) and therefore have a larger probability of reaching thermal energy is taken into account, the β_i and C_i would have to be modified. However, the general form of equation 7-4 remains the same.

† In addition to the "absolute" unit (*i.e.*, percent $\delta k_e / k_e$).

In other words, for a large δk_{ep}, the response is violently fast,* as is the case in an atomic bomb.

b) $\rho \ll \beta$, then $T \to \infty$ and $\lambda_i T \gg 1$. Equation 7-4 becomes

$$T = l^*/\delta k_e$$

and

$$n(t) = n_0 \exp\left(+ \frac{\delta k_e}{l^*} t \right)$$

In other words, for a small δk_e, the response is gentle and depends on the delayed neutrons (through $l^* = l + \beta\bar{\tau}$). This is the condition under which a reactor operates.

In passing, it is interesting to note that when ρ fluctuates around 0 near steady state, T fluctuates at $+\infty$ and $-\infty$, but not going through 0. When ρ is negative and large (reactor scram), the period is determined by τ_6 and the rate of shutdown of a reactor is limited by this effect.

3. One-delayed-neutron-group Approximation

The solution of the inhour equation can be greatly simplified by replacing the six delayed neutron groups with a single average group. In such a case, the inhour equation becomes, for $k_e \approx 1$,

$$\rho = \frac{l}{T} + \frac{\beta}{1 + \lambda T}$$

Solving for T and neglecting λl in comparison with $(\beta - \rho)$, one has

$$T_1 \approx \frac{\beta - \rho + l\lambda}{\lambda\rho} \approx \frac{\beta - \rho}{\lambda\rho} \tag{7-5}$$

$$T_2 \approx -\left(\frac{l}{\beta - \rho + l\lambda} \right) \approx \frac{-l}{\beta - \rho} \tag{7-6}$$

where

$$n(t) = A_1 e^{t/T_1} + A_2 e^{t/T_2}$$

and

$$C(t) = \frac{\beta}{l} \frac{A_1}{1/T_1 + \lambda} e^{t/T_1} + \frac{\beta}{l} \frac{A_2}{1/T_2 + \lambda} e^{t/T_2}$$

The initial conditions

$$\begin{cases} n(0) = n_0 \\ C(0) = \dfrac{\beta}{\lambda l} n_0 \end{cases}$$

---

* When $\rho = \beta$, a reactor is said to be "prompt critical".

yield the equations

$$\begin{cases} n_0 = A_1 + A_2 \\ n_0 = \dfrac{\lambda}{1/T_1 + \lambda} A_1 + \dfrac{\lambda}{1/T_2 + \lambda} A_2 \end{cases}$$

and one has

$$\begin{cases} A_1 = -\dfrac{1 + \lambda T_1}{\lambda(T_2 - T_1)} n_0 \approx \dfrac{\beta}{\beta - \rho} n_0 \\ A_2 = \dfrac{1 + \lambda T_2}{\lambda(T_2 - T_1)} n_0 \approx \dfrac{-\rho}{\beta - \rho} n_0 \end{cases}$$

As an example, consider U-235 fuel for which $\beta = 0.0065$ and $\lambda = 1/\tau = 1/12.9 = 0.0774$. If one sets $\rho = \beta/4$ and $l = 0.001$ sec., then

$$T_1 = 38.7 \text{ sec}$$
$$T_2 = -0.205 \text{ sec}$$
$$A_1 = +1.33 \, n_0$$
$$A_2 = -0.33 \, n_0$$

and

$$\frac{n}{n_0} = 1.33 \, e^{t/38.7} - 0.33 \, e^{-t/0.205}$$

This is plotted in Figure 7.2. The second term on the right causes the initial

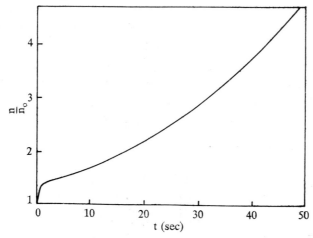

Figure 7.2 Time variation of neutron density due to small step reactivity change, using one-delayed-neutron-group model. ($\beta = 0.0065$, $\lambda = 0.0774$, $\rho = \beta/4$, $l = 0.001$).

"prompt jump" but it dies out quickly. The reactor flux increases eventually with a period of 38.7 second and is mainly due to the delayed neutrons. It is now obvious that although the delayed neutrons account for only a small fraction of the neutrons produced in a reactor, they are of prime importance in reactor kinetics and control.

It should be emphasized that in the derivation of the inhour equation, it is assumed that δk_e is a constant (*i.e.*, a step change in k_e). In the case that δk_e is a function of time (*e.g.*, a ramp function or a harmonic function), the inhour equation would not hold and one must start again from the basic kinetic equations 7-1 and 7-2.

B. Control Rod Theory

To illustrate the effect of a control rod, two cases are considered:

1. *Fully Inserted Rod*

Consider the simple case of a bare homogeneous cylindrical reactor of extrapolated height H and extrapolated radius R. A control rod of radius r_0 is fully inserted along the vertical axis. The problem is therefore a two-region problem and the techniques used in chapter 6 can be applied to get the flux distribution and the criticality condition. If, for simplicity, the one-group modified theory is used,* it is obvious that

$$\frac{d^2\phi}{dr^2} + \frac{1}{r}\frac{d\phi}{dr} + B_R{}^2\phi = 0 \tag{7-7}$$

where

$$B_R{}^2 = B^2 - (\pi/H)^2$$

$$B^2 = \frac{k_\infty - 1}{L^2 + \tau}$$

and $\phi = \phi(r)$, since the axial distribution is that of a cosine function. The solution of equation 7-7 takes the form

$$\phi(r) = A_3 J_0(B_R r) + A_4 Y_0(B_R r) \tag{7-8}$$

where A_3 and A_4 are constants.
The boundary conditions are:

$$\text{a)} \quad \phi(R) = 0$$

$$\text{b)} \quad \frac{d\phi}{dr}\bigg|_{r_0} = \frac{\phi}{r_0 - r_1}\bigg|_{r_0}$$

---

* According to the result of problem 6-b, the theory is simple and yet can be quite accurate for certain applications.

where r_1 is the location inside the control rod, at which the "extrapolated flux" goes to zero. For a black rod (*i.e.*, if it is a perfect thermal neutron absorber), r_1 is related to r_0 through the transport mean free path of the reactor material λ_{tr} as indicated in Figure 7.3.

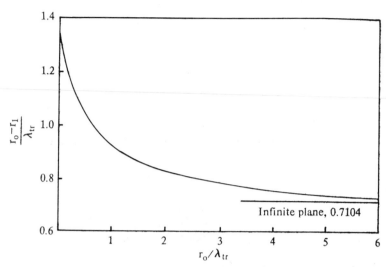

Figure 7.3 Thermal extrapolation distance for black rod. (After Davison and Kushneriuk, MT-214, National Research Council of Canada, Ottawa, 1946).

Using boundary condition (a), one gets from equation 7-8 that

$$A_4 = - \frac{J_0(B_R R)}{Y_0(B_R R)} A_3$$

and thus

$$\phi(r) = A_3 \left[J_0(B_R r) - \frac{J_0 B_R R)}{Y_0(B_R R)} Y_0(B_R r) \right] \qquad (7\text{-}9)$$

where A_3 is proportional to the power level. A qualitative plot of this is given in Figure 7.4 in which the case of no control rod is also plotted for comparison.

The criticality condition can be obtained by applying the two boundary conditions to equation 7-8 and then set the determinant of the coefficients for A_3 and A_4 to zero. This gives

$$\frac{J_0(B_R R)}{Y_0(B_R R)} = \frac{J_0(r_0)}{Y_0(r_0)} \qquad (7\text{-}10)$$

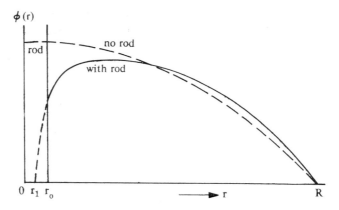

Figure 7.4   Radial flux distribution, with and without black rod.

as the criticality condition. The required fuel concentration given by this equation and the fuel concentration in case of no control rod may lead to a first approximation of the rod reactivity worth.

The theory of off centered control rod and multiple control rods has been developed but is beyond the scope of this text.

2.   *Partially Inserted Rod*

The reactivity worth of a partially inserted rod in a simple system such as that described in the section above can be calculated approximately by such theory as perturbation theory, but the best method is to calibrate the rod experimentally on the reactor itself or on a low-power mockup. Perhaps the most well-known method is the so-called "rod-drop" method. The test rod is withdrawn completely and the reactor is brought to a certain power level and maintain at steady state by means of other control rods. The test rod is then dropped to a pre-determined position, resulting in a negative step reactivity change. By measuring the period (negative in this case) it is possible to determine the reactivity worth of the test rod at that position. If the reactor is now brought back to the same steady power level by means of other control rods, the test rod can be dropped again to yet another position. The result for a system as that in the section above is given in Figure 7.5.

An analytical expression for the curve is, stated here without proof,[*]

$$\frac{\rho}{\rho_{\max}} = \frac{z}{H} - \frac{\sin(2\pi z/H)}{2\pi} \tag{7-11}$$

---

[*] See for example, *Nuclear Reactor Physics* by R. L. Murray, 1957, Page 233.

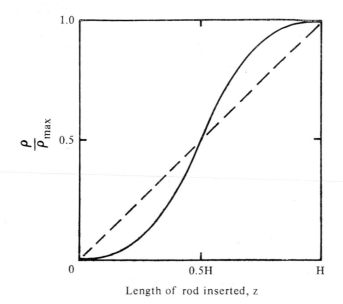

Figure 7.5   Partially inserted control rod worth.

C.   Reactor Control—General Considerations

There are three main factors that make it necessary in the design of a reactor such as a pressurized water reactor to include as much as 20% of excess fuel (and, of course, the necessary control elements) to insure proper operation of the reactor. These effects are discussed below.

1.   *Temperature Effect*

When a power reactor core changes from a cold initial condition to a hot operating condition, there is associated with the temperature increase a volume expansion and an increase in the neutron temperature.

Consider a homogeneous core. The volume expansion coefficient γ is given by

$$\frac{1}{V}\frac{dV}{dT} = \gamma$$

It is easy to see from simple dimensional analysis that since

$$N \propto V^{-1}$$

where N is the nuclei density, and since

$$L \propto N^{-1} \propto V \quad \text{or} \quad L^2 \propto V^2$$

$$\sqrt{\tau} \propto N^{-1} \propto V \quad \text{or} \quad \tau \propto V^2$$

$$B^2 \propto V^{-2/3}$$

one has

$$\frac{1}{L^2} \frac{dL^2}{dT} = 2\gamma$$

$$\frac{1}{\tau} \frac{d\tau}{dT} = 2\gamma$$

$$\frac{1}{B^2} \frac{dB^2}{dT} = -\tfrac{2}{3}\gamma$$

$$\frac{1}{k_\infty} \frac{dk_\infty}{dT} = 0 \qquad (k_\infty \text{ is dimensionless})$$

Thus, since $k_e = k_\infty L_f L_t$, one has

$$\frac{1}{k_e} \frac{dk_e}{dT} = \frac{1}{k_\infty} \frac{dk_\infty}{dT} + \frac{1}{L_f} \frac{dL_f}{dT} + \frac{1}{L_t} \frac{dL_t}{dT}$$

Recalling that $L_f = \exp(-B^2\tau)$ and $L_t = 1/(1 + B^2L^2)$, it is easy to show that

$$\frac{1}{k_e} \frac{dk_e}{dT} = -\tfrac{4}{3}\gamma \left(B^2\tau + \frac{B^2L^2}{1 + B^2L^2} \right)$$

Since γ and k_e are positive, it is obvious that dk_e/dT is negative.* The factor $(1/k_e)(dk_e/dT)$ is called the temperature coefficient of a reactor.

Volume expansion also results in a decrease in the macroscopic cross-section due to the decrease in nuclei density. Furthermore, the increase in neutron temperature leads to, in general, a decrease in microscopic cross-sections. The total effect of the volume expansion and decrease in cross-sections is indicated by the fact that as much as 7% of excess reactivity is required to maintain criticality of a pressurized water reactor (Yankee). For comparison, a safety rod is normally worth about 5%.

---

* This is true for homogeneous reactors and also most heterogeneous reactors, but dk_e/dT may be positive for some heterogeneous reactors such as the OMRE (organic moderated and cooled).

There is one more factor associated with temperature change. Due to the relative speeds of neutrons and nuclei at different temperatures, the resonance absorption cross-sections are different from those of a cold core. This effect, called the Doppler broadening, results in a broader and flatter resonance peak at higher temperature and both the fission and absorption rate increase. The resonance-escape probability usually become lower and, in general, k_e decreases. Excess reactivity in the order of 2.5% (Yankee) is necessary to maintain criticality.

In passing, it is worthwhile to note that pulsed reactors such as the FLASH use the self-limiting charactcristics of the negative temperature coefficicnt and the large neutron leakage factor to terminate a pulse.

2. Fission Product Poisoning

Certain fission products and members of their decay chains have high neutron absorption cross-sections and act as poison to the neutron economy of a reactor. The two most poisonous (high yield and high absorption cross-sections) are Xe-135 and Sm-149. The decay chains involved are:

$$I^{135} \xrightarrow{6.7h} Xe^{135} \xrightarrow{9.2h} Cs^{135} \text{ (practically stable)}$$

$$Nd^{149} \xrightarrow{2h} Pm^{149} \xrightarrow{53h} Sm^{149} \text{ (stable)}$$

The fission yields for thermal fission of U-235 are approximately:

$$I^{135} = 5.6\%*$$
$$Xe^{135} = 0.3\%$$
$$Nd^{149} = 1.1\%$$
$$Pm^{149} \text{ and } Sm^{149} = 0\%$$

The thermal neutron absorption cross-sections of the poison are approximately:

$$Xe^{135} = 3 \times 10^6 \text{ barns}$$
$$Sm^{149} = 5 \times 10^4 \text{ barns}$$

It is thus obvious that Xe-135 is more poisonous than Sm-149. Figure 1.1 shows the Xe-135 concentration as a function of time after a neutron burst.

Due to their high absorption cross-sections, these poisons are "burned up" during reactor operation, but Xe-135 reaches a maximum after reactor shutdown. It is therefore essential that a reactor has available excess reactivity to over-ride this poison and insure reactor startup again without having had

---

* Actually, Te-135 has a very high fission yield but it decays to I-135 with a half-life of less than 1 minute. The fission yield given here is the sum of that for Te-135 and I-135.

to wait for the Xe-135 to die down.* Since Sm-149 is stable, it builds up monotonically after reactor shutdown. As an illustration, the Shippingport pressurized water reactor has a reactivity provision of 3.1% for equilibrium Xe-135 during operation, 2.9% for peak Xe-135, and 0.7% for equilibrium Sm-149. (The excess reactivity for the temperature coefficient is, however, only 2.6% for that reactor.)

It is interesting to note that poison is not always bad. Engineers are using "burnable poison" for reactor control.

3. *Fuel Depletion*

Fuel is consumed during reactor operation. Since it is time consuming and expensive to shut down a reactor to replace the spent fuel frequently, as much as 10% of excess reactivity is sometimes installed at initial start-up (Shippingport). This amount is dependent on the conversion ratio, defined as the number of fissionable nuclei formed per fuel nucleus destroyed. Burnable poison can be best applied with the fuel so that it is not necessary to have too many control rods.

For an uniformly loaded reactor, *i.e.*, if Σ_f is constant in space, the power density is highest at the center of the core where the flux is highest. To get the maximum possible amount of power out of a reactor, it is desirable to have a constant power density (in space). This can be achieved by increasing the fuel concentration at places where the neutron flux is low. This is, however, complicated by the fact that even if the power density is constant everywhere, it is not easy, if possible at all, to remove the heat at a constant rate everywhere. Furthermore, for this type of fuel loading, the neutrons are not utilized most efficiently since the flux is lowest where the reactivity is highest.

It is obvious that by using different methods of fuel loading and shuffling and by using different designs of the heat-removal system, such as using coolant passages of different sizes and orifices, forcing the coolant to flow at different speeds at different channels, or directing the flow at different geometric patterns, an optimum design may be obtained for a particular reactor core such that the highest thermal efficiency can be obtained. A good design should, of course, take into account cost and other engineering operational parameters.

---

* During the Manhattan Project, the Hanford reactor first experienced this difficulty. After an automatic shutdown due to the poison and after the Xe-135 decayed away, the reactor started up again by itself. The reader is referred to an excellent book, *Manhattan Project* by S. Groueff, 1967, in which many important engineering problems are discussed.

SOLVED PROBLEMS

7-1 Derive an expression for t_s, the slowing-down time.

Answer: Consider a neutron with speed v, lethargy u, and scattering mean free path λ_s. Since the time between collisions is λ_s/v and the average change in lethargy between collisions is ξ, it is obvious that

$$\frac{du}{\xi} = \frac{dt}{\lambda_s/v}$$

Since $du = -2\, dv/v$, one has

$$t_s = \int^t dt = \int_{v_f}^{v_t} \frac{-2\lambda_s\, dv}{\xi v^2}$$

where v_f and v_t are the speeds of fission and thermal neutrons, respectively. If an average scattering mean free path $\bar{\lambda}_s$ is assumed, one has, since $v_f \gg v_t$,

$$t_s = \frac{2\bar{\lambda}_s}{\xi v_t}$$

7-2 Develop formulae for the kinetic response of reactor power caused by a small step change in k_e.

Answer: The kinetic response of reactor power P is a function of temperature change T (and, therefore, of time t). Since power is proportional to neutron density, one has, in case of $\rho \ll \beta$,

$$\frac{dP}{dt} = \frac{\delta k_e}{l^*} P$$

for a step change in k_e. The power does not increase exponentially since k_e is a function of temperature and the latter changes with time.

If energy is extracted from a reactor at a time rate of P_e, one has

$$\frac{dT}{dt} = \frac{P - P_e}{c} \tag{7-13}$$

where $T = T(t)$, $P = P(t)$, $P_e = P_e(t)$, and $c = $ heat capacity of the core, generally a constant.

If $(-a)$ is the temperature coefficient of the core and $k_e \approx 1$, then

$$-a = \frac{1}{k_e}\frac{dk_e}{dT} \approx \frac{dk_e}{dT}$$

or,

$$k_e = k_{e0} - aT$$

$$k_e - 1 = k_{e0} - 1 - aT$$

$$\delta k_e = \delta k_{e0} - aT \tag{7-14}$$

where k_{eo} is the effective multiplication factor at time $t = 0$ (or when the temperature change $T = 0$).

For example, consider the case that $P_e = 0$. Eliminating t from equations 7-12 and 7-13, one has an expression for dP/dT and one can solve for $P(T)$. Equations 7-12 or 7-13 then gives $P(t)$ and $T(t)$.

7-3 Consider a bare homogeneous cylindrical reactor of height H and a control rod guide along the vertical axis. A black control rod of length $\frac{1}{2}H$ is placed at the center of the reactor in one case and at the bottom in another case. How much more effective is the rod in the middle?

Answer:

$$\frac{\rho(\text{middle})}{\rho(\text{bottom})} = \frac{\left[\dfrac{0.75H}{H} - \dfrac{\sin\left(2\pi\dfrac{0.75H}{H}\right)}{2\pi}\right] - \left[\dfrac{0.25H}{H} - \dfrac{\sin\left(2\pi\dfrac{0.25H}{H}\right)}{2\pi}\right]}{0.5}$$

$$= \frac{0.818}{0.5} = 1.64$$

SUPPLEMENTARY PROBLEMS

7-a Starting from the kinetic equations 7-1 and 7-2, derive a differential equation for $n(t)$ for a ramp change of $\delta k_e = gt$, where g is a constant, using the one-delayed-neutron-group model.

7-b Using the one-delayed-neutron-group model, $l = 0.001$ sec, and U-235 fuel, plot ρ vs. T for T between 0.001 and 1000 sec.

7-c Obtain explicit expressions of $P(t)$, $T(t)$, and $\delta k_e(t)$ for the example given in problem 7-2. Give a qualitative plot of these functions.

7-d Give qualitative plots of $P(t)$, $T(t)$, and $\delta k_e(t)$ for the case where P_e is proportional to T. Explain briefly the shapes of the curves. Do not attempt to solve any equation. (The differential equation for $T(t)$ is non-linear.)

Chapter 8

HEAT TRANSFER AND FLUID FLOW

A. Heat Removal—General Considerations

THE PROBLEM of heat removal from a reactor core is of prime importance to a nuclear engineer because:

1. Heat is the main product of a power reactor and is certainly a product of any reactor.
2. The coolant used in a reactor often acts as a moderator or reflector, thereby affecting the basic design of a reactor core.
3. The effectiveness of heat removal affects the temperature of the core directly and thus has a definite effect on the effective multiplication factor.
4. The designs of the ex-core coolant loops, heat exchangers, condensers, pumps, generators, etc., strongly depend on the heat generation rate of a reactor and the coolant used. These ex-core equipments may amount to a very large portion of the capital cost of a power reactor plant.
5. It is essential that the temperature inside a core be such that there is no danger of any part of the core getting too hot and melting.

In the study of heat transfer, the important factors are:

1. The physical properties of the coolant such as density, heat capacity, viscosity, thermal conductivity, etc., as a function of temperature and pressure.
2. The geometry of the coolant channel, often given by the hydraulic diameter or equivalent hydraulic diameter.
3. The flow pattern of the coolant and the speed of the coolant flow.
4. The change of phase of the coolant.
5. The heat transfer coefficient, h, at any surface of interest. In fact, most problems of heat transfer amount to the determination of h under the different physical conditions of 1 through 4 above.

B. Conduction, Convection, and Radiation

Basically, the problem of heat conduction in a solid is governed by the heat diffusion equation or Fourier's equation

$$\frac{\partial T}{\partial t} = \frac{k}{\rho c} \nabla^2 T + \frac{1}{\rho c} q \tag{8-1}$$

where q is the heat generated per unit volume per unit time (BTU/ft³-hr), T is the temperature field (°F) in space (r) and time (t), ρ is the density of the material (lb/ft³), c is the heat capacity of the material (BTU/lb-°F), and the thermal conductivity k (BTU/hr-°F-ft), is, by definition, given by

$$Q = -kA \frac{dT}{dr} \tag{8-2}$$

where Q is the amount of heat conducted (loss) per unit time (BTU/hr) across an unit area A (ft²) perpendicular to the direction r where a temperature gradient dT/dr exists. At steady state, the heat generated inside a boundary must be equal to the heat loss outward across that boundary.

It is easy to show (see problem 8-1) that for a hollow cylinder of inside radius r_i, outside radius r_0, and length H,

$$q = \frac{2\pi k H \, \Delta T}{\ln (r_0/r_i)} \tag{8-3}$$

where ΔT is the temperature difference across the cylinder walls.

Conduction involves macroscopic motion of a fluid and heat transfer by convection is usually related to the transmission of heat across a solid-fluid interface. According to Newton's law of cooling, one has

$$Q = hA(T_s - T_c) \tag{8-4}$$

where T_s is the temperature of the solid at the interface, T_c is the average adjacent temperature of the coolant, and h is the heat transfer coefficient (BTU/hr-ft²-°F).

Another method of heat transfer is by radiation. This is important only when the coolant is gaseous. The basic equations governing radiative heat transfer are

$$Q_r = A_1 \epsilon \sigma (T_1^4 - T_2^4) \tag{8-5}$$

$$Q_r = h_r A_1 (T_1 - T_2) \tag{8-6}$$

where T_1 and T_2 are the absolute temperatures of the hotter and colder bodies, respectively. (The absolute zero is at -460 °F). A_1 is the area of the hotter body, σ is called the Stefan's constant and is equal to 0.173×10^{-8} BTU/hr-ft²-°F⁴. The dimensionless factor ϵ is related to the emissivities and absorptivities of the two bodies and Q_r is the net amount of radiative heat transfer rate (BTU/hr). Thus, the radiative heat transfer coefficient

$$h_r = \epsilon \sigma \frac{T_1^4 - T_2^4}{T_1 - T_2} = \epsilon \sigma (T_1 + T_2)(T_1^2 + T_2^2) \tag{8-7}$$

C. Determination of Heat Transfer Coefficient, h

The heat transfer coefficient is a function of many variables. In this section, working formulas for the calculations of h are presented. It is, however, necessary to first define certain quantities commonly used in hydrodynamics:

v = average coolant speed (ft/hr)

H = height of coolant channel (ft)

D_e = equivalent hydraulic diameter of coolant (ft)

$\quad$ = 4 times flow area/wetted perimeter

D_0 = outside diameter of annular coolant (ft)

D_i = inside diameter of annular coolant (ft)

c_c = heat capacity of the coolant (BTU/lb-°F)

k_c = thermal conductivity of the coolant (BTU/hr-°F-ft)

ρ_c = density of the coolant (lb/ft^3)

g = gravitational acceleration constant = 32.2 ft/sec^2.

μ = viscosity of the coolant (lb-mass/ft-sec)

$\quad$ 1 Bvu (British viscosity unit) = 6.72×10^{-4} centipoise

$\quad$ 1 poise = 1 dyne-sec/cm^2 = 1 gram/cm-sec.

v = kinematic viscosity = μ/ρ_c

$[Re]$ = Reynolds number = $D_e v \rho_c / \mu$ (dimensionless)

$[Pr]$ = Prandtl number = $c_c \mu / k_c$ (dimensionless)

$[Nu]$ = Nusselt number = $h D_e / k_c$ (dimensionless)

$[G]$ = Graetz number = $(\pi D_e / 4H)[Re][Pr]$ (dimensionless)

$[Pe]$ = Peclet number = $D_e v \rho_c c_c / k_c = [Re][Pr]$ (dimensionless)

Laminar flow = when $[Re] \lesssim 2{,}320$

Turbulent flow = when for most practical cases, $[Re] \gtrsim 2{,}320$

Working formulas for $[Nu]$, from which the heat transfer coefficient h can be found, are available in almost any standard text in heat transfer and are the results of numerous empirical efforts. Some of these formulas are extracted and presented below, with the understanding that they may be used to get rough answers. For more accurate and modified formulas to suit various situations, the readers are referred to other texts on heat transfer. (*E.g., Fundamentals of Heat Transfer*, by Grober, Erk, and Grigull, 1961, McGraw-Hill Book Co.)

1. Non-metallic fluid; turbulent forced convection; circular pipe of diameter D; valid for H/D ranges from 10 to 400:

$$[Nu] = 0.032[Re]^{0.8}[Pr]^n (D/H)^{0.054} \tag{8-8}$$

where $n = 0.37$ for a heated liquid and 0.3 for a cooled liquid.

$$[Nu] = 0.024[Re]^{0.8}[Pr]^n \text{ for } H/D = 200 \tag{8-9}$$

2. Water and air flowing in annuli; turbulent forced convection; heat transfer across the inner surface only; valid for D_0/D_i ranges from 1.65 to 17:

$$[Nu] = 0.02(D_0/D_i)^{0.53}[Re]^{0.8}[Pr]^{1/3} \tag{8-10}$$

3. Non-metallic fluid; forced laminar flow in a long, straight pipe:

$$[Nu] = 2.02[G]^{1/3}(\mu_f/\mu_w)^{0.14} \tag{8-11}$$

where μ_f = viscosity of bulk fluid and μ_w = viscosity of fluid near wall.

4. Non-metallic fluid; free convection; vertical plate:

$$[Nu] = 0.55[G]^{1/4}[Pr]^{1/4} \quad \text{for} \quad 1700 < [G]^{1/4}[Pr]^{1/4} < 10^8 \tag{8-12}$$

$$[Nu] = 0.13[G]^{1/4}[Pr]^{1/4} \quad \text{for} \quad [G]^{1/4}[Pr]^{1/4} > 10^8 \tag{8-13}$$

5. Air, hydrogen, carbon dioxide, water, or viscous oil in vertical annular space; free convection:

$$k_a/k_c = 0.11[G]^{0.29}[Pr]^{0.29} \quad \text{for} \quad 6000 < [G][Pr] < 10^6 \tag{8-14}$$

$$k_a/k_c = 0.40[G]^{0.20}[Pr]^{0.20} \quad \text{for} \quad 10^6 < [G][Pr] < 10^8 \tag{8-15}$$

where the apparent thermal conductivity k_a is given by (see equation 8-3)

$$q = \frac{2\pi k_a H \,\Delta T}{\ln(D_0/D_i)}$$

and k_c is the true thermal conductivity of the coolant.

6. Liquid metal conduction; long tube with constant Q/A:

$$[Nu] = 0.625[Pe]^{0.4} \quad \text{for} \quad 50 < [Pe] < 20,000 \tag{8-16}$$

7. Same as above, but for heating through both walls of a long flat channel:

$$[Nu] = 10.5 + 0.036[Pe]^{0.8} \tag{8-17}$$

D. Temperature Distributions

Consider a long cylindrical and vertical heterogeneous cell with fuel element of length H and radius r_s. For convenience, the cladding will be neglected. A coolant is assumed to be immediately outside the fuel and has outside radius r_c. The origin is placed at the bottom of the fuel element instead of at the center so that the coolant enters the channel $z = 0$. A moderator may be outside of the coolant, but the conductive and convective heat transfer from the coolant to the moderator is generally neglected. Define

$T_f(0, z)$ = temperature of the fuel at $r = 0$
$T_f(r_s, z)$ = temperature of the fuel at the fuel surface
$T_c(z)$ = temperature of the coolant at (r, z) for $r_s < r < r_c$.

In this section, expressions for $T_f(0, z)$, $T_f(r_s, z)$, and $T_c(z)$ are sought.

1. Axial Temperature Distribution of the Coolant

Since the flux distribution in the axial direction is of the form $\sin (\pi z/H)$, the heat generation rate $q(\text{BTU/hr-ft}^3)$ is also of that form, assuming uniform fuel loading. For a heterogeneous cell in which the radial dimension is small in comparison with the channel height H, it will be assumed that q is independent of r. Thus,

$$q(z) = q_H \sin (\pi z/H) \tag{8-18}$$

where q_H is the value of $q(z)$ at $z = H/2$, half way up the channel.

Since the mass of the coolant that passes by an unit length of heated surface is $M = \rho_c v A_c$, where $A_c = \pi(D_0{}^2 - D_i{}^2)/4$ is the cross-sectional area of the coolant, the coolant temperature rise per unit length is given by

$$\frac{dT_c(z)}{dz} = \frac{q(z)(\pi r_s{}^2)}{c_c M} \tag{8-19}$$

or, integrating from 0 to z,

$$T_c(z) - T_c(0) = \frac{H r_s{}^2 q_H}{c_c M} \left(1 - \cos \frac{\pi z}{H}\right) \tag{8-20}$$

The rate of total heat produced by the fuel and removed by the coolant in the channel is given by

$$Q_T = \pi r_s{}^2 \int_0^H q(z)\, dz \quad (BTU/hr)$$

$$= 2H r_s{}^2 q_H \tag{8-21}$$

Thus, equation 8-20 becomes

$$T_c(z) = T_c(0) + \frac{Q_T}{2 c_c M} \left(1 - \cos \frac{\pi z}{H}\right) \tag{8-22}$$

The coolant temperature thus increases monotonically as z increases.

2. Radial Temperature Distribution of the Fuel

Consider a solid cylindrical fuel element located at z of unit length and radius r. At steady state, the heat generated inside this volume must be equal to that conducted away from its surface. Thus, according to equation 8-2,

$$q(z)\pi r^2 = -k_f(2\pi r)\frac{\partial T_f(r, z)}{\partial r}$$

Integrating from 0 to r and assuming that k_f is constant, one has

$$T_f(r, z) = T_f(0, z) - \frac{q(z)r^2}{4k_f} \tag{8-23}$$

Applying equations 8-18 and 8-21, equation 8-23 becomes

$$T_f(r, z) = T_f(0, z) - \frac{Q_T r^2}{8Hk_f r_s^2} \sin\frac{\pi z}{H_t} \tag{8-24}$$

In other words, the radial fuel temperature distribution decreases parabolically at any given z.

3. *Axial Temperature Distribution of the Fuel*

At the surface of the fuel, where $r = r_s$, equation 8-24 gives

$$T_f(r_s, z) = T_f(0, z) - \frac{Q_T}{8Hk_f} \sin\frac{\pi z}{H} \tag{8-25}$$

and one may note that the temperature drop between the surface and the center of the fuel is independent of the radius of the fuel slug.

From equation 8-4,

$$Q(z) = hA[T_f(r_s, z) - T_c(z)] \tag{8-26}$$

If one considers a small section of the fuel of thickness dz at z and radius r_s, then, by definitions of $Q(z)$ and A,

$$Q(z) = q(z)\pi r_s^2 \, dz$$

$$A = 2\pi r_s \, dz$$

and thus equation 8-26 becomes

$$T_f(r_s, z) - T_c(z) = \frac{Q_T}{4Hr_s h} \sin\frac{\pi z}{H} \tag{8-27}$$

Substituting the value of $T_c(z)$ from equation 8-22 into equation 8-27, the result is

$$T_f(r_s, z) = T_c(0) + Q_T\left[\frac{1}{4Hr_s h}\sin\frac{\pi z}{H} + \frac{1}{2c_c M}\left(1 - \cos\frac{\pi z}{H}\right)\right] \tag{8-28}$$

It is seen that $T_f(r_s, z)$ is always higher than $T_c(z)$ except at the inlet and outlet where they are equal. At these two locations, the heat generation rate is zero, according to the assumptions of sinusoidal flux distribution and uniform fuel loading.

The temperature distribution along the central axis can be obtained by substituting equation 8-28 into equation 8-25. This gives

$$T_f(0, z) = T_c(0) + Q_T \left[\left(\frac{1}{4Hr_sh} + \frac{1}{8Hk_f} \right) \sin \frac{\pi z}{H} + \frac{1}{2c_cM} \left(1 - \cos \frac{\pi z}{H} \right) \right]$$

(8-29)

4. Maximum Fuel Temperatures

The hottest point in the heterogeneous cell $(0, z_0)$ can be located by setting $dT_f(0, z)/dz = 0$. This leads to

$$\tan \left(\frac{\pi z_0}{H} \right) = - \frac{c_c M}{2H} \left(\frac{1}{r_s h} + \frac{1}{2k_f} \right)$$

(8-30)

The negative sign indicates that the maximum occurs where $\pi/2 < \pi z_0/H < \pi$, or, where $z_0 > H/2$. Since $\tan \theta$ can have any values between 0 and $-\infty$ for $\pi/2 < \theta < \pi$, there is no doubt that there exists a z_0 such that equation 8-30 is satisfied. In other words, there is always a maximum temperature between the center of the cell and the outlet, along the central axis.

Similar reasoning and analysis of equation 8-28 indicate that $T_f(r_s, z)$ also has a maximum at (r_s, z_s) such that

$$\tan \left(\frac{\pi z_s}{H} \right) = - \frac{c_c M}{2H} \left(\frac{1}{r_s h} \right)$$

(8-31)

and $z_s > H/2$.

It is immediately seen that $z_s > z_0$. The maxima do not occur at the outlet because, basically, the assumed heat generation rate is sinusoidal in the z direction.

The maximum power of a reactor is limited by the melting points of the reactor materials. This can happen at $T_f(0, z_0)$ or at a point where $T_f(r_s, z) - T_c(z)$ is too large (See next section about burnout heat flux).

Equations 8-22, 8-28, 8-29, and their differences are plotted in Figure 8.1. The data used in the plot are given in solved problem 8-3.

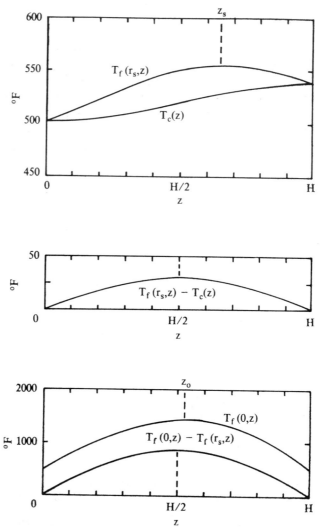

Figure 8.1 Axial temperature distributions of a long cylindrical heterogeneous cell with coolant around fuel rod and uniform fuel loading.

E.   Boiling and Critical Heat Flux

Figure 8.2 is a plot of the heat flux Q/A (BTU/hr-ft^2) and the heat transfer coefficient h (BTU/hr-ft^2-°F) vs. the temperature difference ΔT at the solid-liquid interface for water boiling at atmospheric pressure. The same general characteristics apply to various bulk water temperatures and pressures although the exact location of the peak and the values of h and Q/A at the peaks depend on such parameters as pressure, temperature, surface material and roughness, surface curvature, flow speed, impurity concentration of the fluid, fluid density and viscosity, etc.

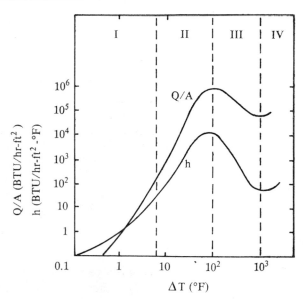

Figure 8.2 Variations of heat flux and heat transfer coefficient with surface-liquid temperature difference for water boiling on a Chromel C wire under atmospheric pressure in a heated pool.

To understand the basic physical mechanisms involved, ΔT is separated into four regions. In region I, there is simply *single phase convection* and no bubbles are formed. In region II, the temperature at the heated surface exceeds the temperature of the saturated vapor (steam) at that pressure and vapor bubbles form at the heated surface. The formation of these bubbles depends on the impurities in the liquid. (These may be solid particles or gas absorbed in the liquid or released from the surface.) In other words, the bubbles are formed at the nucleation sites available. The boiling is often called *nucleate boiling*. The mixing of these bubbles with the liquid reduces the friction at the immediate vicinity of the surface and convection becomes easier in this region. As a result, the heat transfer coefficient increases rapidly. In region III, the bubble formation is so dense that the bubbles coalesce and form a saturated steam vapor film over part of the surface. Since the thermal conductivity of steam is only about one twenty-seventh that of water, the heat transfer coefficient falls rapidly. In this region, the boiling is often called *partial film boiling*. In region IV, an entire region of the heated surface is constantly covered by a stable film. The heated surface usually gets so hot that radiative transfer through the film becomes important and the heat transfer coefficient again rises. In this region, it is said that there is *complete film boiling*.

It is of utmost importance to note that once a peak is reached between regions II and III, the temperature difference at the solid-liquid interface can increase tremendously for the same peak heat flux and the surface gets so hot that melting usually results. The peak heat flux is often called critical heat flux or burnout heat flux. A typical value is 10^6 BTU/hr-ft^2. In the design of nuclear reactors, care must be taken not to exceed the critical ΔT.

Boiling water reactors operate in region II to take advantage of the large heat transfer coefficient. It is indeed unfortunate that the threat of burnout necessitates all power reactors to be operated at lower power levels than they are capable of operating at.

F.   Fluid Flow

The problem of fluid flow in a reactor core is a complicated one. For various reactors, one may have not only those problems as boiling and two-phase flow, but also the problems of radioactive coolant and liquid metal coolant, unimportant for fossil steam plants. For organic coolant, radiation creates gaseous products, tars, and sludge, complicating the operation of the flow system. A detailed study of fluid dynamics is beyond the scope of this text and it may be sufficient to study here the general characteristics of friction, viscosity, pressure drop due to friction, and the effects of sudden enlargement and contraction in the channel size.

1.   *Frictional Pressure Drop*

The semiempirical formula for the pressure drop Δp due to friction in a pipe of length L and diameter D_e is given by the Fanning equation:

$$\Delta p = 4f\frac{L}{D_e}\frac{\rho v^2}{2g}$$ (8-32)

where Δp is given in lb/ft^2 and
  v = average flow speed (ft/sec)
  ρ = density of fluid (lb/ft^3)
  g = 32.2 ft/sec^2
  f = Fanning friction factor (dimensionless).
The friction factor is given by

$$f = 16/[Re] \quad \text{(for laminar flow)}$$ (8-33)

and

$$f = 0.08[Re]^{-1/4} \quad \text{(for turbulent flow in a smooth pipe)}$$

(8-34)

It should be noted that equation 8-33 is for a straight pipe of circular cross-section. For non-circular channels, equivalent diameter may not be used for laminar flow. The roughness of a pipe is often given by the roughness constant ϵ. For a smooth pipe, $\epsilon/D < 5 \times 10^{-6}$ (dimensionless). For commercial pipes, $\epsilon/D \approx 5 \times 10^{-4}$. The Fanning friction factor is plotted against the Reynolds number in Figure 8.3 for various ϵ/D.

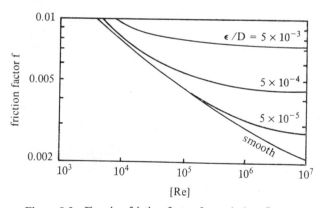

Figure 8.3 Fanning friction factor for turbulent flow.

2. Abrupt Changes in Flow Areas

When a fluid experiences a sudden expansion (due to an abrupt increase in channel cross-section), there is a pressure loss of

$$\Delta p_e = \frac{\rho(v_1 - v_2)^2}{2g} \quad \text{(expansion)}$$ (8-35)

where v_1 and v_2 are the up-stream and down-stream flow speeds, respectively, and ρ and g are defined earlier. For a sudden contraction (due to an abrupt decrease in channel cross-section), the Borda-Carnot expression for the pressure loss is

$$\Delta p_c = K \frac{\rho v_2{}^2}{2g} \qquad \text{(contraction)} \qquad (8\text{-}36)$$

where v_2 is the down-stream flow speed and K can be obtained from Figure 8.4. The up-stream flow speed is v_1 and D_1 and D_2 are the diameters corresponding to v_1 and v_2, respectively.

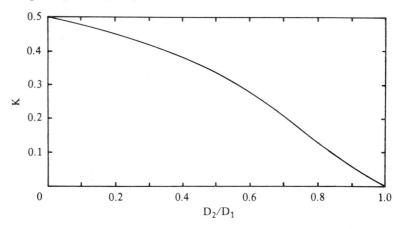

Figure 8.4   Sudden-contraction pressure loss constant.

G.   Thermal Constants of Reactor Materials

Figures 8.5 through 8.10 contain data that may be useful in the study of heat transfer and fluid flow in a reactor.

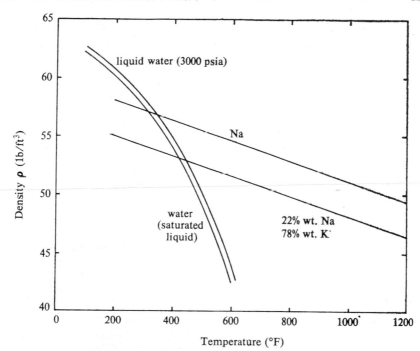

Figure 8.5 Densities of reactor coolant.

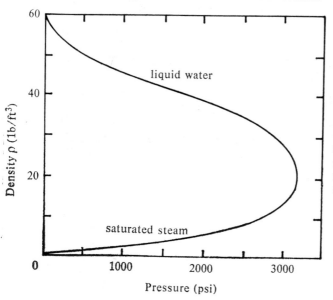

Figure 8.6 Density of steam and water at saturated steam temperature for pressures from atmospheric to critical.

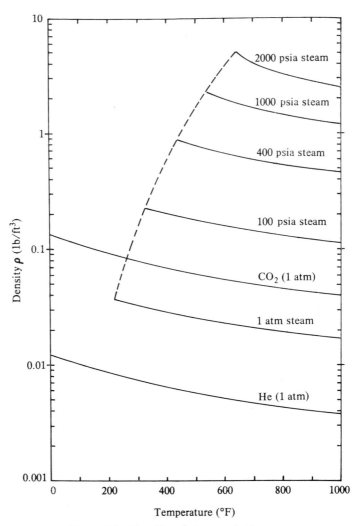

Figure 8.7   Densities of steam and other gases.

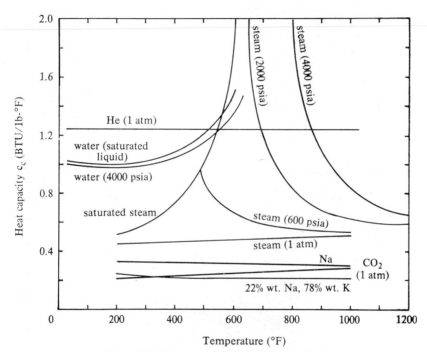

Figure 8.8   Heat capacities of reactor coolants.

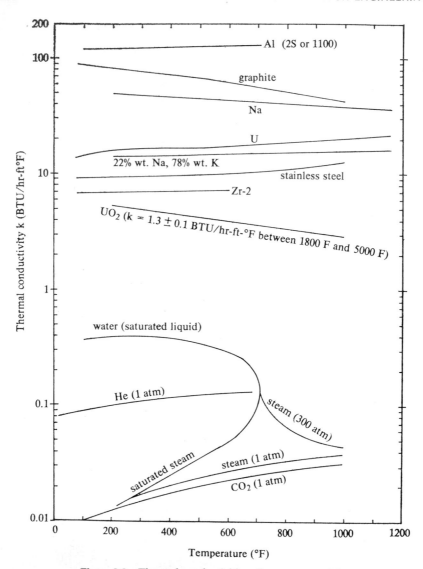

Figure 8.9 Thermal conductivities of reactor materials.

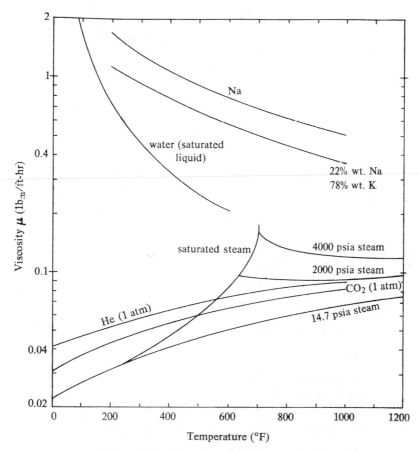

Figure 8.10 Viscosities of reactor coolants.

SOLVED PROBLEMS

8-1 Obtain expressions of Q for fuel of annulus, hollow spherical, and slab geometry, given k_f, ΔT, and the geometric dimensions.

Answer: Consider an annulus fuel element of inside radius r_i and outside radius r_0. Assume that $T_0 > T_i$. Then,

$$\frac{dT}{Q} = -\frac{dr}{k_f A} = \frac{-dr/(2\pi r H)}{k_f}$$

$$\frac{T_0 - T_i}{Q} = -\frac{1}{2\pi k_f H}\int_{r_0}^{r_i}\frac{dr}{r} = \frac{1}{2\pi k_f H}\ln(r_i/r_0)$$

$$Q = \frac{(T_0 - T_i)2\pi k_f H}{\ln(r_0/r_i)}$$

For an hollow sphere of inside radius r_i, outside radius r_0 and $T_0 > T_i$,

$$\frac{T_0 - T_i}{Q} = \frac{-\int_{r_0}^{r_i}\frac{dr}{4\pi r^2}}{k_f}$$

$$Q = 4\pi k_f(T_0 - T_i)/(1/r_i - 1/r_0)$$

For a slab of thickness $x_2 - x_1$ and $T_1 > T_2$,

$$\frac{T_1 - T_2}{Q} = -\frac{x_1 - x_2}{k_f A} \quad \text{or,} \quad Q = k_f A\frac{T_1 - T_2}{x_2 - x_1}$$

8-2 Find expressions of D_e for circular channels, annulus channels, and parallel plate channels.

Answer:

$$D_e = \frac{4\text{ times flow area}}{\text{wetted perimeter}}$$

For circular channel of diameter D, $D_e = 4(\pi D^2/4)/(\pi D) = D$.
For annulus channel, $D_e = 4(\pi D_0^2/4 - \pi D_i^2/4)/(\pi D_0 + \pi D_i) = D_0 - D_i$.
For slab channel of width w and gap g, $D_e = 4wg/(2w + 2g)$.
If $w \gg g$, then $D_e = 2g$.

8-3 A certain pressurized water reactor (PWR) has about 3.5 million UO_2 pellets of 3.4% enrichment. Each pellet is 0.147" in radius and 0.6" high. Twenty five of these pellets form a tube and six of these tubes, stacked ends to ends, form a cell of 7.5' high. A fuel element contains $304 = 16 \times 19$ of these cells and the cylindrical reactor core has 76 elements, each 7.5' high. Each tube has a stainless steel cladding of 0.298" i.d. such that there is only a clearance of 0.002" with the fuel when cold. The cladding is 0.021" thick. The spacings between the cells are 0.422" centers. The coolant is light water, under a pressure of about 2000 psia. There is no phase change. Suppose that

$k_f = 1.3$ BTU/hr-ft-°F (for the oxide at some average fuel temperature)
$\rho_c = 48.7$ lb/ft³ (for some average temperature and pressure)
$T_c(0) = 500°F$ $v = 14$ ft/sec
$c_c = 1.20$ BTU/lb-°F $q_H = 3 \times 10^7$ BTU/ft³-hr
$h = 6000$ BTU/hr-ft²-°F
Study the temperature distributions and power of the cell.

Answer: The flow channel of a cell has an internal diameter $D_i = 0.298'' + 2 \times 0.021'' = 0.34''$. The equivalent flow cross-sectional area $A_c = 0.422^2 - \pi D_i^2/4 = 0.000606$ ft². The equivalent (in flow area) outside diameter D_0 is given by $0.422^2 = \pi D_0^2/4$. That is, $D_0 = 0.476''$.

$$M = \rho_c v A_c = 48.7 \times 14 \times 3600 \times 0.000606 = 1488 \text{ lbs/hr.}$$

$$Q_T = 2Hr_s^2 q_H = 2 \times 7.5 \times (0.147/12)^2 \times 3 \times 10^7 = 6.78 \times 10^4 \text{ BTU/hr.}$$

The maximum heat flux ϕ_m can be obtained by considering an unit length of the cell at the center of the cell. Thus, $2\pi r_s \phi_m = \pi r_s^2 q_H$. One has, therefore, $\phi_m = 1.846 \times 10^5$ BTU/ft²-hr, since $r_s = 0.147''$.

Equations 8-22, 8-28, and 8-29 can now be written as

$$T_c(z) = 500 + 19 \left(1 - \cos \frac{\pi z}{H}\right)$$

$$T_f(r_s, z) = T_c(z) + 30.7 \sin \frac{\pi z}{H}$$

$$T_f(0, z) = T_f(r_s, z) + 870 \sin \frac{\pi z}{H}$$

where $H = 7.5'$. These three equations are plotted in Figure 8.1.

Since there are 304 cells in each fuel element and there are 76 elements in the core, the reactor power is

$$P = 76 \times 304 \times 6.78 \times 10^4 = 1.565 \times 10^9 \text{ BTU/hr} = 458 \text{ Mwt.}$$

Here, Mwt means "megawatt-thermal", since one is talking about thermal energy here and not electrical energy. The readers may note that the reactor discussed here is very close to that of the Rowe Yankee reactor. Modern power reactors are usually designed for about 1000 Mwe. It should be emphasized that the calculations performed here are for the purpose of illustration only. For the actual reactor, the effect of reflectors must be considered. The heat generation rate is actually not constant in the radial direction for the core. The maximum heat flux at the center of the core may be about 2.5 times that calculated earlier. The maximum fuel temperature, for example, is therefore much higher for the core.

8-4 For the flow channel described in problem 8-3 above, what is the approximate pressure loss due to friction? Assume that $\epsilon/D = 5 \times 10^{-4}$ and use the circular pipe model.

Answer: The flow channel of a cell has a cross-sectional area of 0.000606 ft². A circular pipe having the same cross-sectional area would have a diameter $D = 0.333''$. Thus, $[\text{Re}] = (0.333/12)(14 \times 3600)(48.7)/0.22 = 3.09 \times 10^5$.

Figure 8.3 gives $f = 0.005$ and equation 8-32 gives

$$\Delta p = 4 \times 0.005 \frac{7.5}{0.333/12} \frac{48.7 (14)^2}{2 \times 32.2} = 800 \text{ lb/ft}^2 = 5.55 \text{ psi}$$

It may be noted that if one uses $D_e = 4$ times flow area/wetted perimeter $= 4A_c/(\pi D_i) = 0.33''$, the pressure loss calculated would be about the same as the value given above. Of course, one should not use $D_e = D_0 - D_i$ here, since there is no outside wall to cause frictional loss.

K

8-5   For the flow channel described in problem 8-3 above, what is the pressure required to overcome the gravitational potential?

Answer: Recalling that a column of water at room temperature ($\rho = 62.4$ lb/ft^3) and 33.9 ft in height would have a net pressure of 1 atm (14.7 psi) at the bottom, the required pressure is thus

$$p_g = 14.7 \frac{7.5}{33.9} \frac{48.7}{62.4} = 2.54 \text{ psi}$$

8-6   For the flow channel described in problem 8-3 above, if one neglects the effects of the end plates and support plates of the core and assumes that the entrance and exit are sharp, what are the pressure losses due to sudden contraction and expansion? Again, use circular pipe model.

Answer: Consider a cell. The equivalent diameter of the cell is, as stated in problem 8-3, 0.476″. The equivalent cross-sectional area of a cell is 0.001237 ft^2. The flow channel has an area of 0.000606 ft^2 and the corresponding flow diameter is 0.333″, assuming circular pipe model.

Figure 8.4 then gives $K = 0.22$. The pressure loss due to sudden contraction is thus

$$\Delta p_c = 0.22 \frac{48.7(14)^2}{2 \times 32.2} = 32.6 \text{ lb/ft}^2 = 0.226 \text{ psi}$$

To calculate the pressure loss due to sudden expansion, note that for incompressible fluid the velocities of interest are related to the cross sectional areas. Thus, the down stream velocity is

$$v_2 = v_1 \frac{A_1}{A_2} = 14 \frac{0.000606}{0.001237} = 6.858 \text{ ft/sec}$$

and

$$\Delta p_e = \frac{48.7(14 - 6.858)^2}{2 \times 32.2} = 38.57 \text{ lb/ft}^2 = 0.268 \text{ psi}$$

It should be noted that the value of K, as given in Figure 8.4, is defined only for circular pipe. In general, the effects of the end plates and support plates are much more important.

8-7   Considering only the pressures discussed in problems 8-4, 8-5, and 8-6 above, what is the total pumping power required?

Answer: The pumping power is equal to the product of the pressure and the total volume of fluid transported per hour.

The rate of fluid transport for the core is

$$R = 304 \times 76 \times 0.000606 \times 14 = 196 \text{ ft}^3/\text{sec}.$$

The total pumping power is

$$P = (5.55 + 2.54 + 0.226 + 0.268) \times 144 \times 196 = 2.42 \times 10^5 \text{ ft-lb/sec} = 0.328 \text{ Mw}.$$

It is of interest to note that the total pressure drop across the Yankee reactor vessel is about 33 psi. The total pumping power required is therefore only a small portion of the reactor power.

8-8 In problem 8-3, the heat transfer coefficient h is assumed to be 6000 BTU/hr-ft²-°F. Use equation 8-8 to check this value.

Answer:

$$[Pr] = \frac{1.2 \times 0.22}{0.32} = 0.825$$

$[Re] = 3.09 \times 10^5$ ($D = 0.333''$, problem 8-4)

Equation 8-8 gives

$$\frac{h(0.333/12)}{0.32} = 0.032(3.09 \times 10^5)^{0.8}(0.825)^{0.37}\left(\frac{0.333/12}{7.5}\right)^{0.054}$$

$$= 0.032 \times 2.467 \times 10^4 \times 0.9314 \times 0.739$$

$$h = 6250$$

Since h is temperature dependent, the value used is an average value.

SUPPLEMENTARY PROBLEMS

8-a The power densities of some reactors are given below. Convert the units to the British system.

Reactor	Approximate power density (watts/c.c.)
Natural U gas cooled reactor	0.5
Boiling water reactor	30
Pressurized water reactor	55
Sodium cooled fast breeder reactor	760
Nuclear rocket reactor	10,000

8-b Following the discussions of problem 8-3, if the maximum heat generation rate is 2.5 times the average value in the radial direction, what would be the approximate maximum temperatures of the fuel, cladding, and coolant? Assume constant h, k_f, c_c, and μ.

8-c Since the energy generated in the core must be removed by the coolant, check the value (458 Mwt) obtained in problem 8-3.

8-d If the flow speed is doubled in a reactor coolant channel, list ten most important consequences.

Chapter 9

RADIATION HAZARD, PROTECTION GUIDES, AND DETECTION

A. Definitions

1. *Standard Man*

For health physics studies, the "standard man" is over 18 years old. The weights of some of his organs are listed in Table 9.1A and his atomic composition is listed in Table 9.1B. In addition, a standard man is assumed to have a water intake of 2200 c.c./day. He breathes 2×10^7 c.c. of air per day, half during the active 8 hours.

TABLE 9.1A   Weights of organs of the "standard man"

	Weight (grams)
Fat	10000
Skin	2000
Muscle	30000
Bone	7000
Blood	5000
Kidney	300
Liver	1700
Thyroid	20

TABLE 9.1B   Atomic composition of the "standard man"

Elements	Weight (grams)	Weight percentage	Atoms/gram	Atoms percentage
O	45,500	65	2.45×10^{22}	25.7
C	12,600	18	0.903×10^{22}	9.49
H	7,000	10	5.98×10^{22}	62.8
N	2,100	3	0.129×10^{22}	1.36
Ca	1,050	1.5	0.0225×10^{22}	0.236
P	700	1.0	0.0194×10^{22}	0.204
others	550	1.5	0.0161×10^{22}	0.21
	69,500 g (153.2 lb)	100	9.52×10^{22}	100

2. *Curie* (c)

A radioactive source of one curie is the amount of the given radioactive nuclide which undergoes exactly 3.7×10^{10} disintegrations per second. In other words, curie is an unit for activity. The microcurie (μc) unit is often used in radiology. In the case of a radionuclide which undergoes a series of decays, the source strength given represents the parent substance only.

3. *Roentgen* (r)

Roentgen is an unit for radiation exposure. It is the quantity of x- or gamma radiation such that the associated corpuscular emission per 0.001293 gram of air (corresponding to 1 c.c. of dry air at 0°C and 760 mm Hg pressure) produces in air, ions carrying 1 e.s.u. of quantity of electricity of either sign.

In other words, the roentgen is an unit of exposure, measured in energy absorption density in air. Since the charge on an electron is 4.8×10^{-10} e.s.u. and it is accepted that 34.0 ev is required for the formation of one ion pair, one has, in dry air at STP,

$$1r = 34.0/(4.8 \times 10^{-10}) \text{ ev/c.c. of air}$$
$$= 7.07 \times 10^4 \text{ Mev/c.c. of air}$$
$$= 87.7 \text{ ergs/gram of air}$$

For photon energies between 70 kev and several Mev, these values are reasonably energy independent. The corresponding values for soft body tissues, water, and bone are approximately

$$1r = 96.5 \text{ ergs/gram of tissue}$$
$$= 100 \text{ ergs/gram of water}$$
$$= 160 \text{ ergs/gram of bone}$$

4. *Rad*

The rad is an unit of the absorbed dose of radiation. For any material,

$$1 \text{ rad of absorbed dose} = 100 \text{ ergs absorbed/gram of material}$$

5. *Relative Biological Effectiveness* (RBE) *or Quality Factor* (QF)

The RBE is defined by:

RBE = (absorbed dose of Co^{60} gamma radiation to produce a given biological effect)/(absorbed dose of comparison radiation to produce the same effect).

The RBE is related to the linear energy transfer (LET) of the ionizing particle or radiation to water* and is different for different radiation and particles, as shown in Table 9.2. The RBE may also depend on dose, dose rate, presence of oxygen, and physiological conditions. It may be noted that for x-rays with energies between 30 kev and 200 kev, the specific ionization is essentially constant at about 100 ion pairs per micron of water. The RBE is then set at unity.

TABLE 9.2   RBE of various particles and radiation

		RBE	LET (kev/micron of water)	Flux ($\#/cm^2$-sec) to give 2.5 mrem/hr
	thermal	3		670
	100 ev	2		500
	5 kev	2.5		570
	20 kev	5		280
neutrons	0.1 Mev	8		80
	0.5 Mev	10		30
	1.0 Mev	10.5		18
	2.5 Mev	8		20
	5.0 Mev	7		18
	7.5 Mev	7		17
	10 Mev	6.5		17
	30 Mev	5.5		10
x- and gamma rays	10 kev			900
	30 kev			8500
	60 kev			21,000
	0.1 Mev			16,500
		1		
	0.3 Mev			4100
	1 Mev			1300
	3 Mev			570
	10 Mev			230
Electrons and positrons		1		32 (1 Mev)
Protons to 10 Mev		10		0.03 (5 Mev)
Naturally occurring alphas		10		0.0016 (5 Mev)
Heavy ionizing particles		1	3.5 or less	
		2	7	
		5	23	
		10	53	
		20	175	
		30	560	
		35	1000	

* The rate of energy loss per gram from a x-ray is roughly proportional to Z/A between 70 kev and several Mev. For both tissue and water Z/A is about 0.550.

6. *Rem*

The unit of RBE dose or "dose equivalent" is rem, acronym for roentgen equivalent man. *i.e.*,

$$\text{Dose in rems} = (\text{dose in rads}) \times \text{RBE}$$

7. *Mass Absorption Coefficient and Energy Absorption Coefficient*

Consider a flux $\phi(\#/\text{cm}^2\text{-sec})$ of photons, each with energy E (Mev). The rate of energy absorption, not including that contributed by scattering, is $\phi E \mu_a$ (Mev/cm³-sec) where μ_a (cm⁻¹) is the macroscopic absorption cross-section. If $\rho(\text{g/cm}^3)$ is the density of the material and μ_s (cm⁻¹) is the macroscopic scattering cross-section, then $\mu_t = \mu_a + \mu_s$ and, by definitions,

$$\text{mass absorption coefficient } \mu_m = \mu_t/\rho \ (\text{cm}^2/\text{g})$$

$$\text{energy absorption coefficient } \mu_e = (\mu_t - \mu_s)/\rho \ (\text{cm}^2/\text{g})$$

The mass absorption coefficient and energy absorption coefficient of water are plotted in Figure 9.1. The corresponding values for soft tissue and air are about 3% and 12%, respectively, less than that of water.

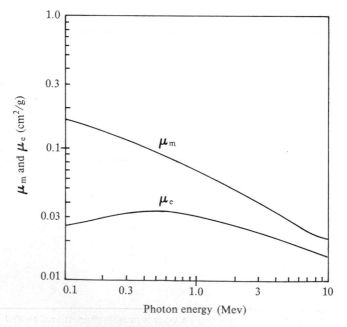

Figure 9.1 Mass absorption coefficient and energy absorption coefficient of water.

B. Biological Considerations

The pertinent biological effects and modifying factors due to radiation are discussed below. These factors are generally interrelated.

1. Biological Variability

Different individual may response to the same dose of radiation in a different manner. An uncertainty factor of four in the dosage required to achieve the same biological effect is not uncommon.

2. Latent Period

There is usually a time delay between the exposure and the manifestation of the biological changes. For small dosages, the delay may be as long as 25 years or more. For large dosage, the delay is short.

3. Recovery and Time Factor

If the dose is not too large, the organism will recuperate. A short burst of radiation usually causes more damage than if the same total dosage is applied over a longer period of time. In other words, the dose rate, as well as total dose, must be considered.

4. Radiosensitivity

Different living organisms are sensitive to radiation in different ways. Virus, for example, can survive radiation about a thousand times stronger than what is lethal to man. Various body cells, tissues, and organs in a mammal differ greatly in radiosensitivity. Young and rapidly dividing cells are most sensitive and fully differentiated nerve cells are most resistant to radiation. Different radioisotopes, when taken internally, produce different levels of toxicity to different tissues and organs. Furthermore, if a radioisotope has a half-life T, the time it takes for the body to get rid of half of the radioisotope (the biological half-life) is evidently not necessarily equal to T.

5. Relative Biological Effectiveness (RBE)

The absorbed dose (in rad) that a body receives may have different effects on the body, depending on the type of radiation and the equivalent RBE dose (in rem).

6. Genetic Effect

Ionizing radiation are capable of producing changes in individual genes and chromosomes in all nucleated body cells. The changes are linearly proportional to the total dose and are generally independent of the duration of the exposure. Sterility, stillborn, and abnormal children may be produced by overexposure to radiation.

C. Occasional and Accidental Exposures

The effect of acute exposure on man is uncertain due to a lack of pertinent knowledge and statistical data. Table 9.3 contains a list of probable effects of acute whole-body external radiation dose received within a twenty-four hour period by an adult. It should be noted that some of the information are derived from experiments with small animals and the effects on man are subject to all probable errors on assumptions, interputations, and extrapolations. Table 9.4 provides the magnitudes of some common radiation sources of interests.

TABLE 9.3   Probable effects of acute whole-body external radiation dose received within a twenty-four hour period by an adult

Total dose (rems)	Probable effects
25	No detectable clinical effects; genetic effect much smaller than the spontaneous rate of mutations; delayed effect unlikely.
50	Slight transient reductions in lymphocytes and neutrophils (*i.e.*, slight blood change); no other clinically detectable effects; radiation induced genetic abnormalities is about the same or smaller than the spontaneously occurring abnormalities; delayed effect possible, but very improbable.
100	Nausea and fatigue; slight blood changes with delayed recovery; genetic effect comparable to that of spontaneous mutations; delayed radiation effect may shorten the life expectancy of man by no more than 1%.
150	Vomiting may occur for a few within 3 hours; fatigue and reduced vitality.
200	Nausea, vomiting, and fatigue will probably occur in most persons within 24 hours; a small portion may die within 2 to 6 weeks; definite depression of practically all blood elements, with recovery taking as long as 6 months; temporary sterility in some cases and possibly permanent sterility in rare instances; delayed effects of major consequence may occur in a small percentage of the individuals.
600	Same as above, with immediate disability, many would never recover completely; all individuals receiving 300 rems or more will vomit within 2 hours, loss of hair within 2 weeks, and with severe blood change to the extent that infection may occur.

Table 9.3 (contd.)

400	Fatal to 50%; recovery of the rest may take a year.
1000	Vomiting within 1 hour and fatal to nearly all within 2 weeks; severe blood change, hemorrhage, infection, and loss of hair.

TABLE 9.4   Magnitudes of certain common radiation sources

Source	Approximate exposure dose
color TV set	7.5 mr/1000 hours at 6 ft away
cosmic rays	30 mr/year
air travel at 25,000 ft.	4 mr/hr
average chest x-ray	200 mr
dental x-ray	5 r/film
gastrointestinal tract (GI) exam.	30 r
x-ray examination during pregnancy	50 r
treatment of tumors (highly localized)	3,000 r to 7,000 r

D.   Protection Criteria

The following guides are sometimes accepted by the radiation workers as the maximum permissible dose (MPD) and dose rate of external radiation and maximum permissible concentration (MPC) for radionuclide taken internally in the form of air, water, and food. For an average person not related to the radiation work, the MPD and MPC should not exceed 10% of that set below for the radiation workers.

1.   *Accumulated External Dose*

a)   Whole body and all critical organ such as head and trunk, skin (cancer), active blood-forming organs (leukemia), gonads (fertility), and eyes (cataracts):

MPD $= 5 (N - 18)$ rem/lifetime, where $N > 18$ is age.

$\quad\quad = 3$ rem/any 13 consecutive weeks

$\quad\quad = 0.1$ rem/week

$\quad\quad = 2.5$ mrem/hr (assuming 40 hr/week; see Table 9.2)

b)   Skin of whole body, excluding the eyes:

MPD $=$ double the above.

c)   Hands and forearms, feet, and ankles:

MPD $= 75$ rems/year

$\quad\quad = 25$ rems/any 13 consecutive weeks.

2. *Emergency and Medical Dose*

An accidental or emergency dose of 25 rems to the whole body or a major portion thereof, occurring only once in the lifetime of a person, is permissible and need not be included in (1) above for radiation workers. Likewise, necessary radiation for dental and medical treatment need not be included. In the case of life-and-death emergency, it is considered reasonable to have up to

> 100 rems in a single exposure
>
> 150 rems in one week
>
> 300 rems in one month

3. *Accumulated Internal Dose*

The maximum permissible *average* concentration (MPC) of radionuclides in air and water are determined on the basis of

$$\text{MPC} = 5\,(N - 18) \text{ rem/life time}$$

$$= 15 \text{ rem/year for most individual organs}$$

$$= 30 \text{ rem/year for thyroid or skin}$$

$$= 5 \text{ rem/year for gonads or whole body.}$$

However, the toxicity of radionuclides varies widely. To illustrate this point, Table 9.5 provides information on a few radionuclides of interest.

The relative hazards of radionuclides absorbed into the body are based on their physical and biological half-lives, MPC in water and air, initial retention and distribution in the body, maximum body burden, types and energies of radiations, and critical organs. Table 9.6 gives a list of some radioisotopes, with the amounts of activities considered as low, intermediate, and high level in radiological laboratories.

4. *Unidentified Radionuclides in Air and Water*

Consider the following groups of radionuclides:

group W1: Ra-226, Ra-228

W2: Sr-90, I-129, Pb-210

W3: Po-210, Ra-223, Th-nat

The maximum permissible concentrations of unidentified radionuclides in water, $(\text{MPCU})_w$ for continuous (168 hr/week) occupational exposure are,

TABLE 9-5 Maximum permissible concentrations for continuous occupational exposures to certain radioisotopes*

Isotope	Physical half-life	Types and energies (Mev) of radiation	MPC in water (μc/c.c.)	MPC in air (μc/c.c.)	Maximum body burden (μc)	Critical organ
Na-24	15.0 h	$\beta(1.39)$ $\gamma(1.37, 2.75)$	6×10^{-3}	10^{-6}	**	GI
Co-60	5.24 y	$\beta(0.31)$ $\gamma(1.17, 1.33)$	10^{-3}	3×10^{-7}	**	GI
Sr-90	27.7 y	$\beta(0.545)$	4×10^{-6}	3×10^{-10}	2	bone
Ra-226	1622 y	$\alpha(4.78)$	4×10^{-7}	3×10^{-11}	0.1	bone
U-235	7.1×10^{8} y	$\alpha(4.4)$	8×10^{-4}	5×10^{-10}	**, 0.03	GI, kidney
U-238	4.5×10^{9} y	$\alpha(4.2)$	10^{-3}	7×10^{-11}	**, 5×10^{-3}	GI, kidney
Pu-239	2.4×10^{4} y	$\alpha(5.0)$	10^{-4}	2×10^{-12}	0.04	bone

* For a more complete listing, see, for example, NBS Handbook #52.
** Maximum body burden is not quoted when gastrointestinal tract (GI) is the critical organ.

TABLE 9.6 Radiotoxicity of selected radioisotopes absorbed into the body

Group	Selected isotopes (listed with increasing A)	Approximate activity range (μc) considered as		
		Low	Intermediate	High
Slightly hazardous	Na-24, K-42, Cu-64, Mn-52, As-76, As-77, Kr-85, Hg-197	$<10^3$	$10^3 - 2 \times 10^4$	$>2 \times 10^4$
Moderately hazardous	H-3, C-14, P-32, Na-22, S-35, Cl-36, Mn-54, Fe-59, Co-60, Sr-89, Nb-95, Ru-103, Ru-106, Te-127, Te-129, I-131, Cs-137, Ba-140, La-140, Ce-141, Pr-143, Nd-147, Au-198, Au-199, Hg-203, Hg-205	$<10^2$	$10^2 - 2 \times 10^3$	$>2 \times 10^3$
Very hazardous	Ca-45, Fe-55, Sr-90, Y-91, Zr-95, Ce-144, Pm-147, Bi-210	<10	$10 - 2 \times 10^2$	$>2 \times 10^2$

measured in μc/c.c. of water,

10^{-7} —if no analysis of the water is made

10^{-6} —if group W1 does not exist

7×10^{-6}—if groups W1 and W2 do not exist

2×10^{-5}—if groups W1, W2, and W3 do not exist.

Next, consider the following groups of radionuclides:

group A1: Pa-231, Th-nat, Pu-239, Pu-240, Pu-242, Cf-249

A2: Ac-227, Th-230, Th-232, Pu-238

A3: all alpha-emitting nuclides and Ac-227 (beta-emitter)

A4: Pb-210, Ra-228, Pu-241

A5: Sr-90, I-129, Pa-230, Bk-249

The maximum permissible concentrations of unidentified radionuclides in air, $(MPCU)_a$ for continuous (168 hr/week) occupational exposure are, measured in μc/c.c. of air,

4×10^{-13}—if no analysis of the air is made

7×10^{-13}—if group A1 does not exist

10^{-12} —if groups A1 and A2 do not exist

10^{-11} —if group A3 does not exist

10^{-10} —if groups A3 and A4 do not exist

10^{-9} —if groups A3, A4, and A5 do not exist.

5. *Flux to Dose Rate Conversion*

Taking the RBE factors into consideration, the conversion curves for gamma and neutron fluxes and dose rates are plotted in Figure 9.2.

E. Radiation Detection

The detection of radiation are all based on electronic excitation and ionization, directly or indirectly. Since gammas and neutrons are not charged, secondary charged particles must be generated in one way or the other. Various methods for neutron detection are discussed in Chapter 2. In this section, the basic principles of some of the most commonly used radiation detectors are presented. These detectors are generally good for detecting gammas, alphas, betas, etc., and with some modifications (such as boron

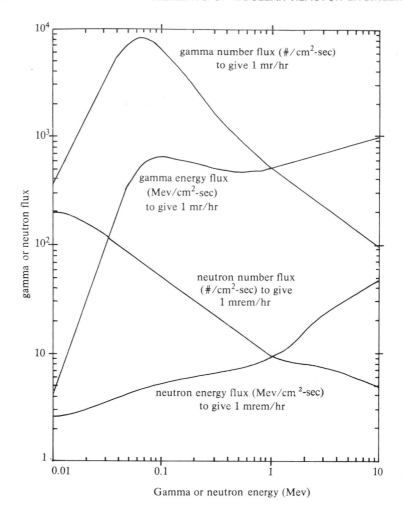

Figure 9.2 Conversion of gamma and neutron fluxes to doses.

linings), can be used to detect neutrons. It is well to note that since charged particles have much smaller range than neutrons or gammas of the same energies, care must be taken to make sure that the walls, if any, of the detector are thin enough for the charged particles to penetrate through. Furthermore, the detection efficiencies and output amplitudes of a detector are usually different for different particles and radiation.

1. *Gas-Filled Detectors*

Perhaps the most well-known detectors are the ionization chamber, the proportional counter, and the Geiger-Muller (G-M) counter. They all have in common a gas-fiilled chamber and a central electrode electrically insulated from the wall. The difference between the three types of detectors is in the voltages applied to the electrodes, the types of gases used, and the number of ions or the magnitudes of the pulse-heights generated by the ionizing particles or radiation directly and indirectly and collected at the electrodes. This is shown in Figure 9.3.

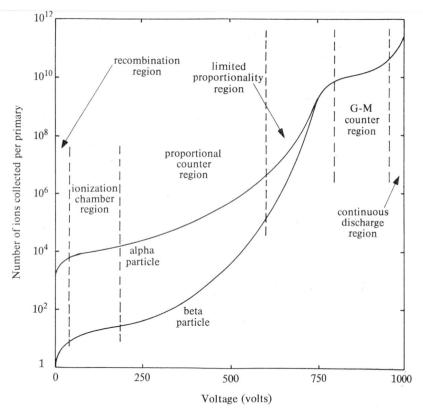

Figure 9.3 Pulse-height versus applied-voltage curves for gas-filled detectors.

Geiger counter For every incoming particle or radiation, it is seen from Figure 9.3 that in the Geiger region approximately 10^{10} ions are generated, mostly through secondary electron collisions near the central electrode. The

L

output pulse height is large and is independent of the types and energies of the incident particle. In other words, the Geiger counter can give information about the number flux, but not energy flux, types of particles, etc.

After each pulse, a G-M counter has a very long dead and recovery time (about 300 microseconds) before it is ready to receive another count again. For this reason, a G-M tube is not accurate for counting rates higher than about 3000 counts per second or for gamma radiation level higher than about 3 mr/hr. Nevertheless, G-M counters are inexpensive, easy to maintain, can be made portable, and are used often for occasional or continuous low level monitoring of air and water.

Ionization chamber, pocket meters, and cutie pie meter   As it is indicated in Figure 9.3, the heights of the signal pulses from an ionization chamber are dependent on the types of particles. When used with appropriate electronic equipments such as pulse-height analyzers (PHA), ionization chambers can be used in the measurements of the specific ionizations and energies of highly ionizing particles such as alphas, as well as betas and gammas. Ionization chambers are used extensively in the field of health physics.

An ionization chamber can also be used as a pocket meter for personnel monitoring. In this case, the instrument is usually made to shape and be worn like a fountain pen. The central electrode, made of graphite coated aluminum, is charged by a small and separate minometer to a known voltage of around 150 volts. The typical capacitance is about $2 \mu\mu f$. After a radiation exposure, the amount of charges, and hence the voltage, on the central electrode decreases by an amount proportional to the total dose received. This change in potential is usually measured by a built-in electrometer in the minometer. The scale is generally calibrated in terms of radiation dosage. A pocket meter is useful in measuring total dose up to around 200 mrad. The accuracy is within 10%, depending on different meters.

Another form of a pocket meter is the self-reading pocket dosimeter, again shaped like a large fountain pen. Although basically still an ionization chamber, it is, however, based on the principle of an electroscope. The fixed and movable loops are made of metal-coated quartz fiber. The small dosimeter is equipped with a lens system and a calibrated scale so that the deflection of the movable loop, when charged, makes an image coincide with zero on the transparent scale when viewed through the eyepiece. Exposure to radiation would, of course, change the separation between the loops and can be read through the calibrated scale anytime without the need of an electronic voltmeter or electrometer.

Still another form of portable ionization chamber is the so-called Cutie Pie (C.P.) meter. Instead of measuring the total dose, however, a C.P. meter is used for dose rate measurements. It is useful in the range of 5 mrad to 20 rads per hour and the accuracy is also about 10%. By using different gases at different pressures inside the tubes, ionization chambers can be used to register still higher dose rates.

There are other forms of ionization chambers, an example of which is the Ohmart cell that make use of the principle of contact potential between the electrodes. It is useful in monitoring the flow of radioactive liquids and in general area monitoring of gammas. A battery is not necessary for the operation of this instrument and a micromicroammeter may be used.

Proportional counter   The most useful mode of operation of a proportional counter is the pulse-type operation. In the proportional counter region (Figure 9.3), the output pulse size is dependent on the specific ionization of the primary radiation and thus information on the types and approximate energies of the primary may be obtained. The large pulse heights (relative to ionization chambers) make possible the use of less expensive electronic equipments. By using a discriminating circuit, it is possible, for example, to count only alpha particles in the presence of beta and gamma radiations, since the alphas give rise to much higher pulse heights.

2.   Scintillation Detector

The use of scintillating phosphors in the detection of radiation dates back to the nineteenth century. Modern scintillators may be in solid, liquid, or even gaseous forms. The most commonly used scintillators are the thallium acti-vated sodium iodide NaI(Tl) crystals and cesium iodide CsI(Tl) crystals for gammas and heavy particles detection and energy determination and anthra-cene and stilbene for electron detection and energy determination.

A scintillator reacts to radiation by emitting a short light pulse of usually less than 25 μsec. This light signal may be collected and processed through a photomultiplier and other electronic circuits. A photomultiplier has a photocathode and many stages of dynodes, each maintained at a higher potential than the previous stage. When the light from a scintillator is picked up by the photocathode, one or more electrons are generated and are acceler-ated toward the dynodes, generating more and more secondary electrons in the process. As a result, an electrical pulse is available as an output from the photomultiplier and may be processed further by such electronic circuits as preamplifier, linear amplifier, pulse discriminator and shaper, pulse height analyzer, and/or scaler.

The short recovery time of a scintillator makes it possible for higher level counting than it is possible with a G-M counter. For each primary particle, the height of the output pulse is a function of the total energy deposited in the crystal, the specific ionization (*i.e.*, the number of ions formed per unit length travelled by the primary), and the geometry and physical characteristics of the system. In conjunction with a pulse height analyzer, the scintillator may be used as an energy spectrometer and various radio-isotopes may be identified by their characteristic locations and relative amplitudes of the energy peaks.

Scintillation detectors can be made portable, but they are more expensive than G-M counters and are not so rugged.

3.   Film Dosimeter

Film dosimeter, or film badges, is based on the same principle as chest x-ray film. Films sensitive to neutrons, gammas, and betas, and shields such as cadmium may be used together and form a small badge to be worn by the radiation workers. The badges are replaced periodically (from 1 week to 3 months) and the films are developed and examined. It is then possible to find out not only what types and how much radiation were received but also the approximate energies of the radiation. The badges are worn by all the workers during working hours while a pocket dosimeter is used inside a more restricted radiation area (such as the vicinity of a reactor beam port) or when an experiment involving radiation is to be performed.

4.   Others

There are many other different types of nuclear radiation detectors, among them are photographic emulsion, Cerenkov detector, chemical dosimeters, calorimeter, cloud chamber, bubble chamber, spark chamber, solid state detectors, atomic absorption and emission spectrophotometers, etc. They are, however, used more often in the area of physics than in reactor engineering and will not be discussed.

5.   Neutrons, Alphas, and Beta Detection

Electrons and alphas are charged and lose their energies quickly when travelling in a medium. As an illustration, Table 9.7 shows some of the values of interest. For the detection and counting of alphas and betas from radio-active decays, it is essential that the effect of self shielding of the isotope and the shielding effect of the counter walls be considered. A 1 Mev electron, for example, has a range of 0.15 cm in aluminum.

For fast neutron detection, recoil protons from the (n, p) reactions are used. For slow neutrons, the (n, α) reactions described in Chapter 2 are used. Most of the detectors described in this chapter may be modified to detect neutrons. For example, it is noted that a BF_3 counter is merely a modified ionization chamber.

TABLE 9.7 Ranges and half-value thicknesses of certain radiation

Radiation	Range in air	Range in water	Thickness of water to stop half of the radiation	Thickness of lead to stop half of the radiation
5 Mev α	4 cm	0.004 cm		
5 Mev β	2600 cm	2.5 cm	0.0004 cm	0.00003 cm
5 Mev γ			9.1 cm	0.6 cm
5 Mev n			1.3 cm	3.3 cm

F. Statistics of Counting

It was mentioned earlier that the dead time of a counter may affect the accuracy of the observed counting rate. This is now investigated further. The dead time of a counter is the time during which the counter is inoperative following a pulse. If a particle enters the counter during this time, it will not be detected. Consider a true signal rate of n cps. Let the observed counting rate be N cps and the dead time be τ sec per pulse. Then the total dead time is $N\tau$ sec in one second and the missing counting rate is $nN\tau$. By equating this to $n - N$, one has

$$n - N = nN\tau$$

or,

$$n = \frac{N}{1 - N\tau} \tag{9-1}$$

It should be noted that the assumption of equally spaced signals has been made. This is, of course, not true, since radioactive decays and background noises such as that due to cosmic rays are random and only statistical in nature. Suppose that in an experiment n counts are obtained in time t, after correcting for the dead time of the counter. Let $\bar{n}$ be the average value of n if many identical experiments are performed. The standard deviation σ and the absolute deviation from the mean ϵ are, by definitions,

$$\sigma^2 = \bar{n}^2 - (\bar{n})^2 \tag{9-2}$$

$$\epsilon = \bar{n} - n| \tag{9-3}$$

For radioactive decay and $\lambda t \ll 1$, it may be shown that

$$\sigma = (\bar{n})^{1/2} \tag{9-4}$$

For large n (say $\geqslant 30$), simple theory of statistics gives the probability $W(n)$ that n counts are obtained in an experiment against an average value of $\bar{n}$ as

$$W(n) = (2\pi\bar{n})^{-1/2} \exp\left[-(\bar{n}-n)^2/(2\bar{n})\right] \qquad (9\text{-}5)$$

or,

$$W(\epsilon) = \frac{1}{\sigma}\left(\frac{2}{\pi}\right)^{1/2} \exp\left[-\epsilon^2/(2\sigma^2)\right] \qquad (9\text{-}6)$$

The function W is often referred to as the Gaussian or normal distribution function. The probability $P(k\sigma)$ of obtaining an absolute deviation from the mean ϵ smaller than $k\sigma$ is obviously

$$P(k\sigma) = \int_0^{k\sigma} W(\epsilon)\, d\epsilon$$

Letting $x = \epsilon/\sigma$,

$$P(k\sigma) = \left(\frac{2}{\pi}\right)^{1/2} \int_0^{k} e^{-x^2/2}\, dx \qquad (9\text{-}7)$$

The function is called the error function and is tabulated in Table 9.8.

TABLE 9.8　Error function

k	Error function P
0.00	0.0000
0.10	0.0796
0.20	0.1586
0.30	0.2358
0.40	0.3108
0.50	0.3830
0.60	0.4516
0.75	0.5468
1.00	0.6826
1.25	0.7888
1.50	0.8664
2.00	0.9546
2.50	0.9876
3.00	0.9974
3.50	0.9996
∞	1.0000

For a few types of errors commonly referred to, Table 9.9 lists the values of k, P, and n required for the given percentage errors. An illustrative example is given in problem 9-7.

TABLE 9.9   Values of k, P, and n for certain types of errors

		Probable error	Standard error	Nine-tenths error
$\dfrac{\lvert \bar{n} - n \rvert}{\sigma} = k =$		0.6745	1.0000	1.6449
$P(k\sigma) =$		0.5000	0.6826	0.9000
Total counts n required for given error	0.1%	4.5×10^5	10^6	2.7×10^6
	0.3%	5.1×10^4	1.1×10^5	3.0×10^5
	1%	4.5×10^3	10^4	2.7×10^4
	3%	506	1.1×10^3	3.0×10^3
	10%	45	100	271

According to Table 9.9, it may be noted that if one takes 3×10^3 counts and calculates the average counting rate, and if one does it many times, then 9 out of 10 times one would get a value within 3% of the true counting rate (obtained if one takes an infinite number of counts).

Whenever absolute counting rate is desired, as in the case of radio-isotope concentration measurements, it is not only necessary to have the basic knowledge of statistics but also the correction factors due to background radiations, counter geometry and back scattering effect, dead time of counter, counter efficiency, etc. In passing, it may be noted that there are criteria to reject some "apparently" inaccurate data. A most frequently used guidance is the so-called Chauvenet's criterion. It states that a measurement in a set of m trials shall be rejected if its deviation from the mean is such that the probability of occurrence of all deviations equally large or larger does not exceed $1/(2m)$. In other words, a measurement is rejected if its k value is larger than

TABLE 9.10   Dependence on m of Chauvenet's limiting k value

m	k
5	1.68
10	1.96
20	2.24
50	2.58
100	2.80
200	3.02
500	3.29

the one given by the relation:

$$1 - P = 1/(2m).$$

Table 9.10 contains a list of different m and the corresponding k at or above which a measurement will be rejected.

SOLVED PROBLEMS

9-1 The maximum body burden is given as 0.04 μc for Pu-239. Check it.
Answer: Pu-239 is an alpha emitter. The maximum body burden is the activity A such that the absorbed dose rate D is equal to 15 rems/yr. Using a RBE factor of 10, D can be converted to

$$D = \left(15 \frac{\text{rems}}{\text{year}}\right)\left(\frac{\text{year}}{365 \times 24 \times 3600 \text{ sec}}\right)\left(\frac{\text{rad}}{10 \text{ rems}}\right)\left(\frac{100 \text{ ergs/gm}}{\text{rad}}\right)\left(\frac{\text{Mev}}{1.6 \times 10^{-6} \text{ erg}}\right)$$

$$= 3 \text{ Mev/g-sec.}$$

Now, Pu-239 is a bone seeker and there are 7000 grams of bone in a standard man. If the Pu-239 is uniformly distributed and the energy per disintegration is E(Mev), then

$$D = \frac{AE}{7000} \text{ (Mev/g-sec)}$$

With $E = 5$ Mev/disintegration,

$$A = \frac{3 \times 7000}{5} = 0.42 \times 10^4 \text{ dis/sec} = 0.11 \ \mu c$$

There is a discrepancy factor of about 2.7 with the listed value. Considering the facts that bone is more sensitive to radiation than other tissue and that the radioisotope is not uniformly distributed in the bone, etc., the values may be considered as in good agreement.

9-2 The maximum body burden is 0.04 μc for Pu-239. How large is a particle with this activity?
Answer:

$$T_{1/2} = 2.4 \times 10^4 \text{ yr} = 7.6 \times 10^{11} \text{ sec} \quad \text{and} \quad \lambda = 1.693/T_{1/2}$$

The number of Pu-239 atoms is equal to the ratio of activity and λ.

$$N = \frac{0.04 \times 10^{-6} \times 3.7 \times 10^{10} \times 7.6 \times 10^{11}}{0.693} = 1.62 \times 10^{15}$$

The weight of the pure Pu-239 particle is related to the Avogadro's number.

$$W = \frac{239 \times 1.62 \times 10^{15}}{6.03 \times 10^{23}} = 6.4 \times 10^{-7} \text{ gram}$$

If the particle is spherical in shape, the diameter is, since the density is 19.7 g/c.c.,

$$d = 4 \times 10^{-3} \text{ cm} = 40 \text{ microns.}$$

This particle, if deposited in the bone as a lump, would cause a bone tumor. Fortunately, a particle of this size is not likely to get through the nasal passages. If swallowed, it most likely will be eliminated because of its low soluability. However, the very hazardous nature of Pu-239 and the need for special handling are obvious.

9-3 Sr-90 is considered as the most hazardous bone-seeking radioisotope because of its high soluability (about 25%), long biological half-life, and strong resemblance to calcium (bone-seeking). Calculate the approximate MPC in water.

Answer:

Let n_w = Sr-90 nuclei density in water ($\#$/c.c.)

V = volume of water consumed per day = 2200 c.c./day

f — fraction of Sr-90 retained = 25%

λ_p = physical decay constant = $0.693/(27.7 \times 365) = 0.7 \times 10^{-4}$ day^{-1}

λ_b = biological decay constant = $0.693/(10 \times 365) = 1.9 \times 10^{-4}$ day^{-1}

M = weight of bone = 7000 grams

N = total number of Sr-90 nuclei in the bone ($\#$)

E = average disintegration energy = $\frac{1}{3}$ maximum beta energy

$\quad$ = 0.33×0.55 Mev/dis.

Since Sr-90 is a beta emitter and the RBE factor is 1, the maximum permissible dose rate D is, similar to problem 9-1,

$$D = 30 \text{ Mev/g-sec.}$$

However,

$$D = N\lambda E/M \text{ (Mev/g-sec)}$$

Thus,

$$N\lambda = \frac{30 \times 7000}{0.33 \times 0.55} \text{ dis/sec}$$

At equilibrium, one has

$$n_w Vf = N(\lambda_p + \lambda_b)$$

The maximum permissible activity A is given by

$$A = \lambda n_w = \lambda N \frac{(\lambda_p + \lambda_b)}{Vf}$$

$$= \frac{30 \times 7000}{0.33 \times 0.55} \frac{(0.7 \times 10^{-4} + 1.9 \times 10^{-4})}{2200 \times 0.25} = 0.54 \text{ dis/sec-c.c.}$$

$$= 14.6 \times 10^{-6} \text{ } \mu c/cc$$

The discrepancy factor of 3.7 with the listed value is due to the facts that bone is more sensitive to beta radiation than other tissue (by a factor of as much as 5) and that the radio-isotope is not uniformly distributed in the bone. In fact, there are "hot spots" in which the Sr-90 concentration may be as much as 60 times higher than other locations.

9-4 What is the exposure dose rate at a point 5 feet from a Co-60 source of 4000 μc?

Answer: Co-60 decays by emitting a 0.31 Mev beta particle, followed by two gammas successively with energies of 1.17 Mev and 1.33 Mev. At the field point of interest, the effect of the betas may be neglected, since the betas have short ranges. The attenuation effect on the gammas due to the air may also be neglected. The gamma number flux at the field point is, due to the inverse-square spreading,

$$\phi = \frac{2 \times 4000 \times 10^{-6} \times 3.7 \times 10^{10}}{4\pi(5 \times 12 \times 2.54)^2} = 1020 \text{ photons/cm}^2\text{-sec.}$$

Since the photons have average energy of 1.25 Mev each, it takes about 420 photons/cm²-sec to give 1 mr/hr, according to Figure 9.2. The exposure dose rate R at the field point is therefore

$$R = \frac{1020}{420} = 2.4 \text{ mr/hr}$$

If one does not use Figure 9-2, one may reason that the rate of energy absorption is

$$\phi E \mu_a = (1020 \text{ photons/cm}^2\text{-sec})(1.25 \text{ Mev/photon})(\mu_e \rho)$$

where $\rho = 0.001293$ g/c.c. and $\mu_e = 0.03$ cm²/g, with the value 0.03 obtained from Figure 9.1. Hence,

$$\phi E \mu_a = 0.0492 \text{ Mev/sec-c.c. of air}$$

Since $1r = 7.07 \times 10^4$ Mev/c.c. of air, the expected exposure dose rate is

$$R = \frac{0.0492}{7.07 \times 10^4} \ r/\text{sec} = 2.5 \text{ mr/hr}$$

It may be noted that the assumption of $1r \approx 1$ rad has been made and no correction factor was applied to change water to air, as in Figure 9.2.

9-5   A counter has a dead time of 200 μsec and 1.8×10^5 counts are observed in 1 minute. What is the true countng rate? If the dead time is 20 μsec instead, what would be the observed counting rate?
Answer: The observed counting rate with the slow counter is 3000 cps. The true counting rate is

$$n = \frac{3000}{1 - 3000 \times 200 \times 10^{-6}} = 7500 \text{ cps}$$

With the fast counter, the observed counting rate N is given by

$$7500 = \frac{N}{1 - N(20 \times 10^{-6})}$$

or, $N = 6520$ cps.

9-6   In a series of experiments in radiation counting, the following 20 counting rates (corrected for dead time, background, etc.) are obtained in 1 minute intervals by twenty students:
320, 287, 315, 299, 289, 312, 307, 296, 285, 285, 314, 312, 300, 292, 297, 295, 306, 304, 284, 387.
Should any point(s) be rejected?
Answer:

$$m = 20. \qquad \bar{n} = 304.25. \qquad (\bar{n})^2 = 92568.0625. \qquad \overline{n^2} = 92732.5.$$

$$\sigma = (92732.5 - 92568.0625)^{1/2} = 12.82.$$

The last point deviates from the mean by

$$387 - 304.25 = 82.75$$

and for this point,

$$k = \frac{|304.25 - 387|}{\sigma} = \frac{82.75}{12.82} = 6.45$$

This is greater than the k value of 2.24 given in Table 9.10. The reading (387) is therefore rejected. Now, consider the remaining 19 readings.

$$m = 19. \quad \bar{n} = 299.89. \quad (\bar{n})^2 = 89936.8. \quad \overline{n^2} = 90090.6. \quad \sigma = 12.40.$$

Of the 19 readings, the first value (320) has the largest deviation from the mean. The corresponding k value is

$$k = \frac{320 - 299.89}{12.40} = 1.62$$

This reading (320) is thus not rejected and the rest will also be kept.

9-7 What is the total number of counts required if one wants a 1% deviation from the true count and with a ninety-five hundredth error?
Answer:

$P(k\sigma) = 0.95$. Table 9.8 gives $k = 1.9600$. Thus, $|\bar{n} - n|/\sigma = 1.96$.

With 1% deviation, $|\bar{n} - n| = 0.01\,\bar{n}$. Since $\sigma = (\bar{n})^{1/2}$, one has

$$\frac{0.01\bar{n}}{(\bar{n})^{1/2}} = 1.96$$

or, $\bar{n} = 3.84 \times 10^4$ counts.

9-8 If the true average for a given time interval is 400, what is the probability of obtaining an absolute deviation of 10 from the mean?
Answer: The probability of obtaining 410 is, according to equation 9-5,

$$W = \left(\frac{1}{2\pi \times 400}\right)^{1/2} \exp\left(-\frac{100}{2 \times 400}\right) = 0.885/50.2 = 0.01765$$

To calculate the probability of obtaining an absolute deviation from 400 of more than 10 counts, note that $\sigma = (400)^{1/2} = 20$. $k = (410 - 400)/20 = 0.5$. Thus, $P = 0.3830$, according to Table 9.8. The desired probability is therefore $1 - 0.3830 = 0.617$.

9-9 Two successive measurements on a single long half-life source, each for the same duration of time and corrected for background radiation, yield $n_1 = 3080$ and $n_2 = 3320$ counts, respectively. Should the mean value be used?
Answer: The effective standard deviation is

$$\sigma = (\sigma_1^2 + \sigma_2^2)^{1/2} = (n_1 + n_2)^{1/2} = 80$$

$$k = (n_2 - n_1)/80 = 3$$

From Table 9.8, $P = 0.9974$. The probability that the fluctuation is statistical in nature is therefore $1 - 0.9974 = 0.26\%$. It is suggested that the variation in counting is due to something else, possibly in the counting equipments. The mean value should not be used without further study.

SUPPLEMENTARY PROBLEMS

9-a Check the maximum body burden value of 2 μc for Sr-90. What is the weight of the Sr-90 dust with this activity?

9-b Check the MPC in air value for A-41. Explain any discrepancy.

9-c Tour a radiation detection laboratory, if available, and draw schematic diagrams and block diagrams of the different detection systems.

9-d In a counting experiment involving C-14, the total counting rate is 20 cpm. If the background counting rate is 13 cpm, how long would it take to measure the C-14 activity to within 4% accuracy?

Chapter 10

SHIELDING

ALTHOUGH a practical and economical reactor shielding design depends largely on the experiences of the engineers, it is nevertheless well worth the effort to look into the problems and be familiar with the basic principles involved. Naturally, the first step is to understand the radiation, such as types, sources, levels, distributions (spatial, energy, and time), interaction with matter (mechanisms and cross-sections), etc. Only then can an attempt be made to determine the necessary materials and thicknesses to reduce the radiation to an acceptable level. In this chapter, these and other related topics are investigated. However, topics such as radiation effect on materials, radiation heating in shields, etc. are omitted.

A. Radiation Sources

The radiation sources in and around a reactor are directly and indirectly due to fission. The chief radiations are fission fragments, alphas, betas, neutrinos, gammas, and neutrons. The first three of these are charged and therefore have very short ranges. However, through nuclear processes such as radioactive decays and bremsstrahlung (to be discussed later), gammas and neutrons may be generated. Neutrinos are extremely penetrating. There is practically no way to stop them, nor is there a necessity to do so, since a direct consequence of their high penetrating power is that they have practically no interaction with the matter through which they traverse and there is no danger of being hurt by neutrinos. The energies carried by the neutrinos are considered as inevitable loss. This leaves gammas and neutrons, both uncharged and penetrating.

1. *Gamma Sources*

The gammas in and around a reactor are from the following nuclear processes, some of them already discussed in Chapter 3.
a) *Prompt fission gammas*—These are the gammas emitted in nuclear fissions.
b) *Decay gammas*—These are the delayed gammas from radioactive decays of fission fragments or activated substances.
c) *Inelastic scattering gammas*—In the (n, n) inelastic scattering processes, the excited nuclei often emit gammas.

d) *Capture gammas*—These are the gammas resulting from neutron captures. Unlike activation gammas, the capture gammas are emitted immediately. An example is the $B^{10}(n, \alpha)Li^7$ reaction. A gamma is emitted from the excited Li^7.

e) *Annihilation gammas*—Activated structural materials may contain Co-58 and Zn-65, etc. These are positron emitters. When a positron meets an electron, two photons of at least 0.51 Mev each are created and go in opposite directions.

f) *Bremsstrahlung gammas*—Whenever energetic electrons are decelerated in an atomic electric field, gammas are created. The process is important when the electrons are highly relativistic (say, over 10 Mev) and the medium has very high Z. However, this process is usually negligible except when Li^7 is used as coolant.

2. *Neutron Sources*

Neutron sources are discussed in detail in Chapter 2. In and around a reactor, however, the following sources are considered to be of prime importance:

a) *Fission prompt neutrons*—These are the neutrons emitted in nuclear fissions.

b) *Delayed neutrons*—These are the neutrons from radioactive decay chains of fission fragments.

c) *Photoneutrons*—Through (γ, n) reactions, energetic neutrons are generated. As far as shielding is concerned, the important nuclei for photoneutron productions are D^2, Be^9, C^{13}, and Li^6, with threshold energies of 2.23 Mev, 1.67 Mev, 4.9 Mev, and 5.3 Mev, respectively.

d) *Activation neutrons*—These are delayed neutrons emitted by activated substances. An important example is the formation of N^{17} through the $O^{17}(n, p)N^{17}$ reaction by fast neutron bombardment of water. (In water, 0.037 % of the oxygen is O^{17}). The N^{17} then beta decays $(T_{1/2} = 4.14 \text{ sec})$ to excited O^{17} and the latter may emit a neutron of about 1 Mev. The average cross-section for the $O^{17}(n, p)N^{17}$ reaction is about 9 μb for fission spectrum neutrons.

B. Interaction of Gammas with Matter

While charged particles lose energies gradually through a large number of small transfers through excitations and ionizations and neutrons lose energies by scatterings, a gamma tends to lose much or all of its energy in a single interaction. (It is, of course, due to this characteristic that one has the well-known exponential attenuation for gammas). There are four main kinds of gamma interactions; namely, photoelectric effect, Compton effect, pair production, and photonuclear reaction. The last of these, an example of which is photoneutron production, is important when the gamma energy is of the order of 8 Mev or more, with the relatively important exception of only the four isotopes mentioned in the previous section on photoneutron. Since most reactor gammas are of energies less than 8 Mev, photonuclear reaction will not be discussed further and the remaining three types of interactions will be investigated. The dominant regions of the interactions are shown in Figure 10.1.

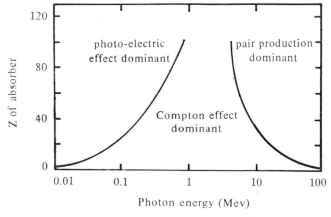

Figure 10.1 Dominant regions for the three main types of gamma interactions. The lines are for the cases where the cross-sections are the same for the neighboring effects.

1.   Photoelectric Effect

At low gamma energies, this effect is dominant. In a photoelectric interaction, there is a complete energy transfer from the photon to a bound atomic electron. The photon simply disappears. The photoelectron has kinetic energy E equal to the difference of the photon energy hf and the binding energy ϕ of the electron in the atom:

$$E = hf - \phi \qquad (10\text{-}1)$$

where f is the frequency of the photon and h is the Planck constant.

For photon energies greater than the K-shell binding energy, about 80% of the photoelectric interactions are with the K-electrons and the total photoelectric cross-section in this energy range is roughly

$$\sigma_{ph} = 10^{-8} Z^n / (hf)^{7/2} \quad \text{barns/atom} \qquad (10\text{-}2)$$

where Z is the atomic number of the atoms, hf is in Mev, and n varies almost linearly from 4.05 to 4.65 as hf increases from 0.1 Mev to 3 Mev.

For photons with energies less than 0.1 Mev, the attenuation coefficient is so high that shielding these photons is generally not the most serious problem. For example, a sheet of lead 2 cm thick or a pool of water $4\frac{1}{2}$ ft deep is more than enough to reduce the soft photon intensity by a factor of 10^{10}.

The angular distribution of photoelectrons is shown in Figure 10.2. The mass attenuation coefficients for gammas in lead are shown in Figure 10.3. It is seen from Figure 10.2 that for 0.5 Mev photons, approximately 50% of the photoelectrons are ejected with an angle smaller than 24° with respect to the primary photons and all photoelectrons may be treated as going in the "forward" direction.

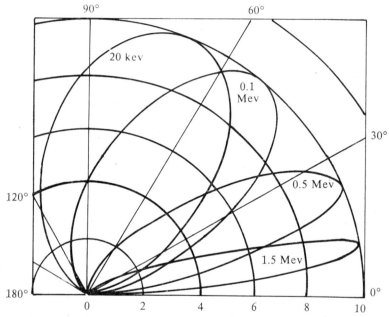

Figure 10.2 Angular distribution of photoelectrons. (Relative number per unit angle, $dn/d\theta$).

2. Compton Effect

Since the rest mass m of an electron is small ($mc^2 = 0.51$ Mev), non-relativistic theory breaks down quickly when electrons with energies above 0.5 Mev are involved. The relativistic theory of photon scattering by a free electron is called Compton scattering. When the binding energy of the struck electron

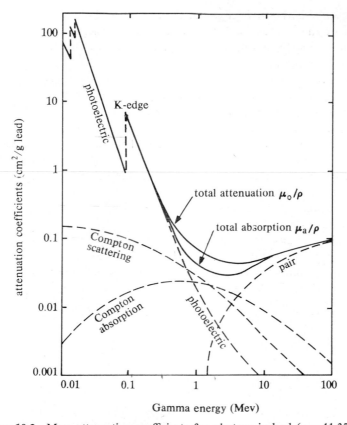

Figure 10.3   Mass attenuation coefficients for photons in lead ($\rho = 11.35$ g/c.c.).

is small in comparison with the photon energy, as is true in the energy range where the photoelectric effect is not prominent, the electron is considered as free and unbound.

Consider the case of a free electron at rest and hit by an incoming photon of energy hf and momentum hf/c. It is simple to show from the laws of conservation of energy and momentum that

$$\lambda' - \lambda = \frac{h}{mc} (1 - \cos \theta) \qquad (10\text{-}3)$$

or

$$\frac{1}{E'} - \frac{1}{E} = \frac{1}{mc^2} (1 - \cos \theta) \qquad (10\text{-}4)$$

M

where $\lambda' = c/f'$ is the wavelength of the scattered photon, θ is the angle of scattering of the photon, and E' and E are the energies of the scattered photon and primary photon, respectively. The energy of the scattered photon hf' is thus determined uniquely by the scattering angle and not by what material it is.

It is very useful to remember that since atomic electrons are considered as free and unbound in the Compton scattering process, it is only the electron density of a material and not the Z of the material that will affect the Compton scattering cross-section. For this reason, the Compton scattering cross-section is generally given units of barns/electron, regardless of what material is involved. It may, however, be noted that the number of electrons in a gram of material does not vary too much for all materials. In other words, the Compton scattering cross-section for a particular photon energy is roughly proportional to the density of the material.

The differential cross-section for the number of photons scattered per electron and per unit scattering angle in the direction θ is given by the well known Klein-Nishina equation (the derivation of which is quantum mechanical and relativistic, Z. Physik, 52, 853, 1929):

$$\frac{d\sigma(E,\theta)}{d\theta} = 2\pi \sin\theta \frac{r_0^2}{2} \frac{1}{[1 + \alpha(1 - \cos\theta)]^2} \left[1 + \cos^2\theta + \frac{\alpha^2(1 - \cos\theta)^2}{1 + \alpha(1 - \cos\theta)} \right]$$

(10-5)

where

$$\alpha = E/(mc^2)$$

$$r_0 = e^2/(mc^2) = 0.282 \times 10^{-12} \text{ cm}$$

$$(4\pi r_0^2 \approx 1 \text{ barn})$$

and the unit of $d\sigma(E, \theta)/d\theta$ is barn/electron-radian. The right hand side of equation 10-5 is plotted in Figure 10.4. For a more complete set of graphs of the Compton energy-angle relationship and the Klein-Nishina formula, see NBS Circular 542 and Gamma-Ray Absorption Coefficients, by C. M. Davisson and R. D. Evans, Rev. of Mod. Phys., 24, 79, 1952.

The cross-section for the number of photons scattered with scattering angle less than θ_1 is given by

$$\sigma(E, \theta_1) = \int_0^{\theta_1} \frac{d\sigma(E, \theta)}{d\theta} d\theta \quad \text{(barn/electron)}$$

(10-6)

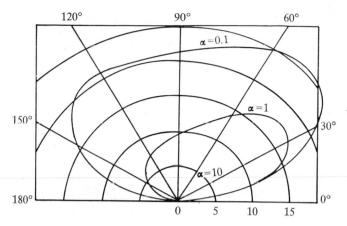

Figure 10.4 Differential cross-section per unit angle for the number of Compton scattered photons in the direction θ (10^{-26} cm^2/electron-radian).

When $\theta_1 = \pi$, this becomes

$$\sigma(E) = 2\pi r_0^2 \left\{ \frac{1 + \alpha}{\alpha^2} \left[\frac{2(1 + \alpha)}{1 + 2\alpha} - \frac{1}{\alpha} \ln(1 + 2\alpha) \right] \right.$$

$$\left. + \frac{1}{2\alpha} \ln(1 + 2\alpha) - \frac{1 + 3\alpha}{(1 + 2\alpha)^2} \right\} \tag{10-7}$$

and is the cross-section of the total number of photons scattered per electron and per incoming photon of energy E. This corresponds to the μ_0/ρ curve in Figure 10.3 and the μ_m curve in Figure 9.1, provided that the conversion factor for electron density is applied. In other words, for a particular material such as water or lead,

$$n\sigma = \mu_0/\rho = \mu_m \tag{10-8}$$

where $\sigma = \sigma(E)$, $\mu_0 = \mu_0(E)$, $\mu_m = \mu_m(E)$, and n is the number of electrons per gram of material.

Similarly, the cross-section for the amount of energy scattered per electron and per unit angle is

$$\frac{d\sigma_e(E, \theta)}{d\theta} = \frac{E'}{E} \frac{d\sigma(E, \theta)}{d\theta} \tag{10-9}$$

and the cross-section $\sigma_e(E)$ for the total amount of energy removed from the initial beam may be obtained by integrating equation 10-9 from $\theta = 0$ to $\theta = \pi$. With respect to μ_a/ρ in Figure 10.3 and μ_e in Figure 9.1, one has, for a particular material,

$$n\sigma_e = \mu_a/\rho = \mu_e \qquad (10\text{-}10)$$

3. Pair Production

When a gamma ray with energy of at least 1.02 Mev passes through the field of a charged particle, such as the vicinity of a heavy nucleus, the gamma may be completely absorbed and a positron-electron pair appears. The process is called pair production. Most of the gamma energy in excess of 1.02 Mev (which is the rest mass equivalent of the positron-electron pair) are shared by the pair in the form of kinetic energies. However, the positron generally carries a larger portion of the total excess energy, especially for low gamma energy. This is due to the nuclear repulsion of the positron and the nuclear attraction of the electron. It is, indeed, not unusual for the positron to have twice as much kinetic energy as the electron. The positron-electron pair usually go in the "forward" direction. The average angle (in radian) between the incident photon and the outgoing electron is of the order of mc^2/T for $T \gg mc^2$, where T is the kinetic energy of the electron. Thus, the tendency to have a forward distribution is more pronounced when the gamma is more energetic.

The cross-section for pair production is complicated, but is generally proportional to Z^2. The energy dependence of the pair production cross-section for lead is plotted in Figure 10.3. For very high energy photons (20 Mev or higher for lead), screening effect due to the inner electrons of the atoms further complicates the cross-section calculations and tends to decrease the cross-section of pair production. In fact, the cross-section per nucleus for pair production is lower for lead than for air or aluminum at high photon energies (up to 25 Mev). However, since reactor gammas are usually not so energetic, the screening effect may well be neglected.

C. Build-up Factor

In shielding, the treatment of uncollided flux (or virgin flux), involving particles or photons that emerge without experiencing any collisions, is generally simply an exponential attenuation calculation. When the effect of collided flux (or secondary flux) is taken into account, however, the exercise becomes, in reality, a difficult transport problem. To get around this complication, a semiempirical build-up factor (BUF) may be used as a multiplication factor to correct for the effect due to scattered flux. There are many types of build-up factors, the most important ones are the number buildup factor B_n, energy buildup factor B_E, energy absorption buildup factor B_a, and dose buildup factor for gamma B_r ,defined below:

$$B_n = \frac{\text{total number flux } (\#/\text{cm}^2\text{-sec}) \text{ at a point}}{\text{uncollided number flux at the same point}}$$

$$= \frac{\int \phi \, dE}{\int \phi^0 \, dE} \tag{10-11}$$

$$B_E = \frac{\text{total energy flux } (\text{Mev}/\text{cm}^2\text{-sec}) \text{ at a point}}{\text{uncollided energy flux at the same point}}$$

$$= \frac{\int \phi E \, dE}{\int \phi^0 E \, dE} \tag{10-12}$$

$$B_a = \frac{\text{total energy absorbed } (\text{Mev}/\text{c.c.-sec}) \text{ at a point}}{\text{energy absorbed due to uncollided flux at the same point}}$$

$$= \frac{\int \mu_0 \phi E \, dE}{\int \mu_0 \phi^0 E \, dE} \tag{10-13}$$

$$B_r = \frac{\text{total dose rate } (\text{r}/\text{hr}) \text{ at a point}}{\text{dose rate due to uncollided flux at the same point}}$$

$$= \frac{\int \mu_0^{\text{air}} \phi E \, dE}{\int \mu_0^{\text{air}} \phi^0 E \, dE} \tag{10-14}$$

where in all four cases,

$$\phi = \phi(r, E) = \text{number flux spectrum} = \frac{\text{number flux}}{\text{unit energy}} \left(\frac{\#/\text{cm}^2\text{-sec}}{\text{Mev}} \right)$$

$\phi^0 = \phi^0(r, E) = $ uncollided number flux spectrum

$\mu_0 = \mu_0(E) = $ linear total attenuation coefficient (cm^{-1}), same as the one in Figure 10.3.

Obviously, the BUF is a function of the photon energy E, shield thickness x, and shield material.

Consider the simple case of number flux variation for gamma attenuation by a shield of thickness x. If only virgin flux is considered,

$$\phi^0(x) = \phi_0 \, e^{-\mu_0 x} \tag{10-15}$$

where $\phi_0 = \phi(0)$ is the flux at $x = 0$. If secondary flux is considered,

$$\phi(x) = B_n \phi^0(x) = B_n \phi_0 \, e^{-\mu_0 x} \tag{10-16}$$

One very useful expression for B_r is

$$B_r = A\, e^{-a\mu_0 x} + (1 - A)\, e^{-b\mu_0 x} \qquad (10\text{-}17)$$

where the energy and material dependent constants A, a, and b can be calculated from experimental data for three different x and are tabulated. To calculate these constants accurately using equation 10.11 requires high speed computer.

Other expressions and data for B_n, B_E, B_a, B_r, etc., are available (See *Reactor Shielding Design Manual* by T. Rockwell, D. Van Nostrand Co.). Representative values are given in Table 10.1 for point isotropic source and for $\mu_0 x$ between 0 and about 15.

TABLE 10.1   Values of the coefficients for the dose BUF

Shield		Gamma energy (Mev)		
		1	**2**	**10**
Water	A	11	6.4	2.7
	a	−0.104	−0.076	−0.042
	b	0.030	0.092	0.13
Ordinary concrete	A	10	6.3	2.6
	a	−0.088	−0.069	−0.050
	b	0.029	0.058	0.084
Iron	A	8	5.5	2.0
	a	−0.089	−0.079	−0.095
	b	0.04	0.07	0.012
Lead	A	2.37	2.6	0.5
	a	−0.043	−0.07	−0.215
	b	0.185	0.08	0.08

D.   The Monte Carlo Method (Random Sampling)

There are many interesting analytical methods in the study of gamma attenuation. With the advent of high speed computer, a very interesting and useful method, the so-called Monte Carlo method, is used on an increasing scale. Basically, the life history of each and every gamma is traced, according to the rules of chance. The important processes involved in gamma attenuation are, in the following order:

1. initial energy
2. initial direction
3. distance to next interaction
4. types of nucleus involved
5. types of interaction
6. energy and angle of scattered photon
7. azimuthal angle
8. polar angle

As an illustration, consider the 5th step in which the exact type of interaction is desired. Suppose that one is interested in only photoelectric effect, Compton scattering, and pair production. The corresponding cross-sections for the given nucleus (step 4 earlier) and given gamma energy (step 1 earlier) are σ_{pe}, σ_c, and σ_{pp}, respectively. If the total cross-section is

$$\sigma_t = \sigma_{pe} + \sigma_c + \sigma_{pp}$$

then the probability of absorption is

$$P_a = \frac{\sigma_{pe} + \sigma_{pp}}{\sigma_t} < 1$$

and the probability of scattering is

$$P_s = \frac{\sigma_c}{\sigma_t} < 1$$

such that $P_a + P_s = 1$.

If one picks a random number R_5 between 0 and 1 and $R_5 < P_a$, one may then consider the interaction is either photoelectric effect or pair production, resulting in the complete absorption of the gamma. (The gamma is dead. Steps 6 through 8 are no longer necessary). If $R_5 > P_a$, the gamma may be considered as being Compton scattered and one may go onto steps 6 through 8 and then cycle from step 1 again. The process for a gamma ends if and only if the gamma is absorbed or escapes the shield.

It is obvious that to get good statistical results, many gammas have to be considered. There are various ingenious methods that would effectively reduce the total number of samplings. One of these interesting methods, for example, has the exotic name of Russian Roulette. However, it is beyond the scope of this text to talk about gambling Russians.

E. Relaxation Length and Removal Cross-Section

A very useful concept in the study of radiation attenuation is in the use of a quantity called relaxation length. It is recognized that the collided flux does not follow the exponential law, due to buildup. However, by defining a quantity s such that

$$e^{-x/s} = B_n e^{-\mu_0 x} \tag{10-18}$$

one can then write the collided flux as

$$\phi = \phi_0 e^{-x/s} \tag{10-19}$$

The quantity s is called relaxation length (cm) and it is a quantity in a simple exponential law that takes into account the effect of buildup. It has been found that for fast neutron attenuation by a thick shield, the relaxation lengths of certain shielding materials are that shown in Table 10.2. The corresponding values for 4 Mev gammas in water and lead are 30 cm and 2.4 cm, respectively.

TABLE 10.2   Relaxation length and removal cross-section for certain shielding materials

	Relaxation length for isotropic point fission neutron source s(cm)	Removal cross-section for fission neutrons in slabs $\Sigma_r(\text{cm}^{-1})$
ordinary concrete	12	*
water	10	0.0978
aluminum	10	0.079
barytes concrete	9.5	*
lead	9	0.12
graphite	9	0.058
beryllium	9	0.128
iron concrete	6.3	*
iron	6	0.17

* Contains water

Another common concept is the so-called removal cross-section. It is useful only for fast neutron attenuation. In most cases, it is determined by using a thin (compared with relaxation length) slab of solid material, a plane fission neutron source on one side, and a thick (at least 4 or 5 relaxation length) layer of water on the other side. By measuring the fast neutron fluxes with and without the solid shield, the cross-section of the solid in removing fast neutrons may be computed. The macroscopic removal cross-sections for various materials are given in Table 10.2. It should be noted that the thin solid shield does not absorb the fast neutrons. It merely slows them down and removes them from the category of fast neutrons. The slow neutrons will be thermalized and absorbed by the water, leaving only those fast neutrons that escape both the shield and the water to be detected.

It has been found that the removal cross-section of a material is constant for shield thickness up to about five relaxation lengths of that material.

If a point source is used, the removal cross-section Σ_r is related to the relaxation length s by

$$\Sigma_r = 1/s \tag{10-20}$$

F. Shielding Materials

All practical shielding materials have common characteristics such as fire resistance, non-toxic, odorless, inexpensive, durable, easy to install, etc. The most important and unusual properties of some of the basic shielding materials are given below:

1. *Iron*

Iron, as carbon steel or stainless steel, is good structural material and is used for thermal shield as well as radiation shields. The density of iron is 7.85 g/c.c. Impurity contents such as cobalt and manganese in steel are the subjects of many studies, since these elements emit induced gammas after shutdown.

2. *Lead*

Lead is an excellent gamma shield, at both the high and low energy ends. At 3 Mev, where the absorption coefficient is at a minimum, it is about as effective as the same mass of iron. Lead does not have objectionable impurities and can withstand radiation without damage. Where space and weight are important, as is the case for mobile reactors or radioisotope sources, lead is a premium choice as gamma shield. However, it has poor structural properties and the low melting point of 618°F may well indicate the fact that the tensile strength of lead at 300°F drops to 37% of that at room temperature and is only about 1% as that of steel at 300°F. The density of lead is 11.35 g/c.c.

3. *Concrete*

There are three main types of concrete; namely, ordinary concrete, Barytes concrete, and iron aggregate concrete. The density of ordinary concrete is about 2.3 g/c.c. The compositions of the other two types of concretes are listed below in Table 10.3.

TABLE 10.3   Barytes and iron aggregate concrete

	Barytes concrete		Iron aggregate concrete	
Composition (wt. %)	Barytes	60	steel punching	57
	Limonite	22	Limonite	26
	Portland cement	11	Portland cement	13
	water	7	water	4
density	3.5 g/c.c.		4.5 g/c.c.	

In choosing which types of concrete to use, it is important to note that the major factors are density and cost per cubic yard installed. A thick shield usually increases the cost of other instrumentation such as the lengths of pipes, etc. Another factor of importance is the structural strength of concrete in a radiation field. It is believed that structural concrete should not be exposed to an energy flux higher than 2×10^{11} Mev/cm²-sec in order that the temperature may not rise appreciably.

4.   *Water*

Because of the high hydrogen density in water, it is a very good neutron shield. It is important, however, to make sure that the water is pure and does not contain dissolved salts that will become radioactive or will help the decomposition of the water into hydrogen and oxygen gas. Water, when used in laminated shield, also serves as the coolant for the shield.

5.   *Others*

Other reactor materials such as graphite and uranium also act as a shield. Figure 10.5 shows the thickness of various materials to attenuate a narrow beam of gammas by a factor of 10. No buildup factor is used.

G.   Geometric Considerations

In this section, a few of the practical and simplest shielding geometry transformations will be studied. Uncollided number flux will be considered, although collided flux may similarly be considered by including the BUF.

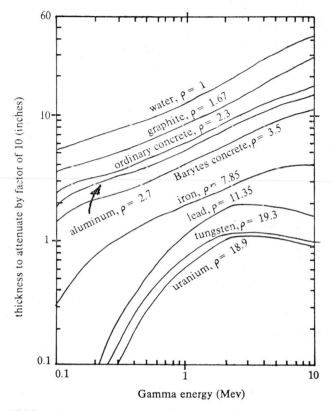

Figure 10.5 Thickness to attenuate a narrow beam of gamma rays by a factor of 10. Densities are in g/c.c.

The analysis are generally good for both gammas and neutrons, using appropriate absorption coefficients, relaxation lengths, or removal cross-sections. In most cases, symmetry are used so that the solutions are of simpler forms. This is considered sufficient since the techniques are adequately demonstrated.

1. Point Kernel

A point source of strength S_0 ($\#$/sec) in an infinite homogeneous medium characterized by μ_0 will have a flux ($\#$/cm² sec) at a distance R given by

$$\phi = S_0 \frac{e^{-\mu_0 R}}{4\pi R^2} \tag{10-21}$$

where the factor $e^{-\mu_0 R}$ is due to material attenuation and the factor $1/(4\pi R^2)$

takes care of the inverse-square spreading. The combined factor

$$G(R) = \frac{e^{-\mu_0 R}}{4\pi R^2} \qquad (10\text{-}22)$$

is often referred to as the point attenuation kernel. The flux due to any geometry may be obtained by integration, at least in principle.

2. *Line Source*

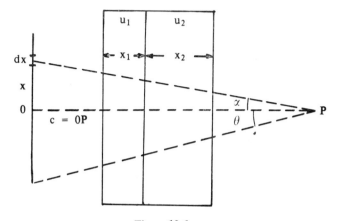

Figure 10.6

Consider an uniform line source S_L (#/cm-sec) and a field point P symmetrically located a distance c away. A laminated shield of total mean free path

$$b = \mu_1 x_1 + \mu_2 x_2 \qquad (10\text{-}23)$$

is between 0 and P. The uncollided flux at P due to the line element dx is

$$d\phi_L = \frac{S_L\, e^{-b \sec \alpha}}{4\pi(c \sec \alpha)^2}\, dx$$

Since $x/c = \tan \alpha$ and $dx = c \sec^2\alpha\, d\alpha$,

$$\phi_L = \frac{S_L}{4\pi c} \int_{-\theta}^{\theta} e^{-b \sec \alpha}\, d\alpha$$

Defining

$$F(\theta, b) = \int_o^\theta e^{-b \sec \alpha} \, d\alpha \qquad (10\text{-}24)$$

one has

$$\phi_L = \frac{S_L}{4\pi c} \left[F(\theta, b) - F(-\theta, b) \right] = \frac{S_L}{2\pi c} F(\theta, b) \qquad (10\text{-}25)$$

The "sec integral" F is tabulated. Some of the useful relationships are:

$$F(-\theta, b) = -F(0, b) \qquad (10\text{-}26a)$$
$$F(0, b) = 0 \qquad (10\text{-}26b)$$
$$F(\theta, 0) = \theta \qquad (10\text{-}26c)$$
$$F(\text{small } \theta, b) = \theta \, e^{-b} \qquad (10\text{-}26d)$$

3. *Disk and Infinite Plane Sources*

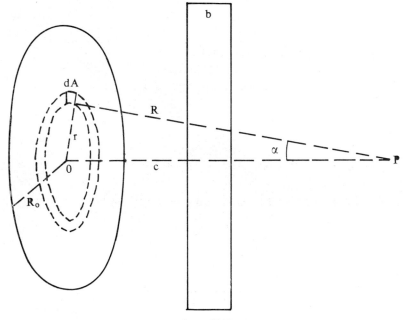

Figure 10.7

Consider a thin disk source of radius R_0 and uniform source surface density S_A ($\#/\text{cm}^2\text{-sec}$). A point P is a distance c from the center 0 along the axis.

A shield characterized by b (see equation 10-23) is between 0 and P. The flux at P due to the disk source is obviously

$$\phi_D = \int G(R)S_A \, dA$$

where

$$G(R) = \frac{e^{-b \sec \alpha}}{4\pi R^2} = \frac{e^{-bR/c}}{4\pi R^2}$$

and

$$\phi_D = \int_0^{R_0} G(R)S_A 2\pi r \, dr.$$

Since $r^2 + c^2 = R^2$, one has $r \, dr = R \, dR$ and

$$\phi_D = 2\pi S_A \int_c^{\sqrt{c^2 + R_0^2}} G(R)R \, dR \tag{10-27}$$

For infinite plane source, $R_0 = \infty$ and

$$\phi_P = 2\pi S_A \int_c^{\infty} G(R)R \, dR$$

or,

$$\phi_P = \frac{S_A}{2} \int_b^{\infty} \frac{e^{-y}}{y} \, dy$$

where $y = bR/c$. Defining an "exponential integral" by

$$E_n(x) = x^{n-1} \int_x^{\infty} \frac{e^{-y}}{y^n} \, dy \tag{10-28}$$

one has

$$\phi_P = \frac{S_A}{2} E_1(b) \tag{10-29}$$

$$\phi_D = \frac{S_A}{2} \left[E_1(b) - E_1\left(b \frac{\sqrt{c^2 + R_0^2}}{c} \right) \right] \tag{10-30}$$

The function $E_n(x)$ is tabulated. Some of the useful relationships are:

$$E_n(\text{large } x) = \frac{e^{-x}}{x} \tag{10-31a}$$

$$E_0(x) = \frac{e^{-x}}{x} \tag{10-31b}$$

$$E_0(0) = E_1(0) = \infty \tag{10-31c}$$

$$E_2(0) = 1 \tag{10-31d}$$

$$E_3(0) = 0.5 \tag{10-31e}$$

$$\int_{[x}^{\infty} E_n(y)\, dy = E_{n+1}(x) \tag{10-31f}$$

4. Spherical Surface Source

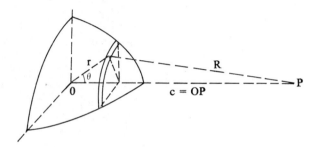

Figure 10.8

Consider a hollow thin sphere with radius r and uniform source surface density S_A ($\#/\text{cm}^2\text{-sec}$). The flux at P due to the sphere is

$$\phi_S = \int G(R)S_A\, dA = \int_0^{\pi} G(R)S_A 2\pi(r \sin \theta)(r\, d\theta)$$

Since $R^2 = r^2 + c^2 - 2rc \cos \theta$ and $2R\, dR = 2rc \sin \theta\, d\theta$, one has

$$\phi_S = 2\pi S_A \frac{r}{c} \int_{c-r}^{c+r} G(R)R\, dR \tag{10-32}$$

$$= \frac{r}{c}\left[\phi_P(c - r) - \phi_P(c + r)\right] \tag{10-33}$$

In other words, a spherical surface source may be replaced by two infinite plane sources suitably located. Furthermore, if equation 10-32 is rewritten as

$$\phi_S = 2\pi S_A \frac{r}{c} \int_{c-r}^{\sqrt{(c-r)^2+4rc}} G(R)R \, dR$$

and is compared to equation 10-27, one has

$$\phi_S = \frac{r}{c} \phi_D(c - r, \sqrt{4rc}) \tag{10-34}$$

i.e., the uncollided flux from a spherical surface source may be replaced by that from a disk located a distance $(c - r)$ from point P and has radius equals $(4rc)^{1/2}$, provided that the self-shielding of the source itself may be neglected.

5.   *Semi-Infinite Slab (Finite Thickness) Source*

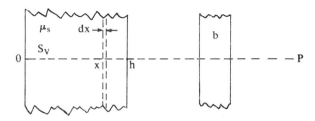

Figure 10.9

Consider a semi-infinite slab of thickness h, uniform volume source density S_V ($\#$/c.c.-sec), and absorption coefficient μ_s for the source. If a shield characterized by b is between the slab and point P and if the slab is considered as being composed of many infinite planes, then the flux at P due to the slab is

$$\phi_I = \int_0^h \frac{S_V}{2} E_1 \, dx$$

where

$$E_1 = E_1[b + \mu_s(h - x)] \tag{10-35}$$

Defining $z = b + \mu_s(h - x)$ and $b' = b + \mu_s h$, one has

$$\phi_I = \frac{S_V}{2\mu_s} \int_b^{b'} E_1(y) \, dy.$$

Since

$$\int_b^{b'} = \int_b^{\infty} - \int_{b'}^{\infty}$$

one has, using equation 10-28f,

$$\phi_I = \frac{S_V}{2\mu_s} [E_2(b) - E_2(b')] \tag{10-36}$$

6. Solid Cylindrical Source

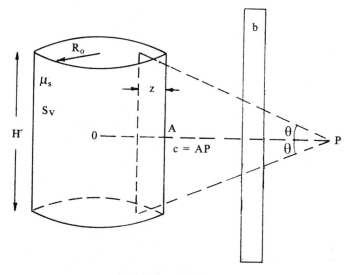

Figure 10.10

An interesting method of handling analytically a solid cylindrical source is to represent the cylinder by an equivalent (same strength) line source. The problem is to find the location characterized by z, as shown in the figure above.

Considering the self-shielding effect of the cylinder and writing

$$S_V \pi R_0^2 H = S_L H$$

$$\phi = \frac{S_L}{2\pi(c + z)} F(\theta, b'')$$

N

where $b'' = b + \mu_s z$, one has

$$\phi = \frac{S_V R_0^2}{2(c + z)} F(\theta, b'') \tag{10-37}$$

The function F is defined by equation 10-24.

7. Solid Sphere Source

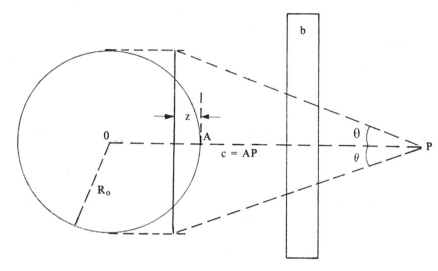

Figure 10.11

A solid sphere source may be transformed into an equivalent disk source of the same diameter. Writing

$$S_V(\tfrac{4}{3}\pi R_0^3) = S_A \pi R_0^2$$

and

$$\phi = \frac{S_A}{2} \left[E_1(b'') - E_1(b'' \sec \theta) \right]$$

where $b'' = b + \mu_s z$, one has

$$\phi = \tfrac{2}{3} R_0 S_V [E_1(b'') - E_1(b'' \sec \theta)] \tag{10-38}$$

H. Rules of Thumb

A set of simple rules are now given for reactor shielding. These rules may be treated as guidelines for rough calculations:

1.   For intermediate energy gammas (0.5 Mev to 5 Mev), the effectiveness of the shield is roughly proportional to the total weight of the material, since Compton scattering is roughly proportional to electron density.

2.   For high and low energy gammas, heavy materials should be used, since pair production is roughly proportional to Z^2 and photoelectric effect is roughly proportional to $Z^{4.3}$.

3.   The relaxation length for isotropic point fission neutron source is about 10 cm for all shielding materials. The removal cross-sections for fission neutrons are higher for heavier materials, because of inelastic collisions. In other words, fast neutrons (higher than 1 Mev) should be shielded by heavy materials.

4.   Hydrogeneous materials should be used to moderate neutrons of intermediate energies (below 1 Mev).

5.   Boron compounds may be added to other shielding materials such as water to enhance neutron absorption without producing gammas.

6.   Concrete may be used to shield both gammas and neutrons, although it is not the most effective for either radiation.

7.   Composite laminated shield is usually the most effective and economical when both neutrons and gammas are to be shielded. Heavy and hydrogeneous materials should be arranged alternatively in order that such secondary radiation as capture gammas and photoneutrons may be shielded.

8.   Sources of different geometries may often be easily transformed to simplier geometries.

9.   The effective gamma energy used in reactor plant shielding may be conservatively assumed to be 1 Mev, although computer codes and multiple energy groups are usually used. The most important exception is the short lived energetic N-16 gammas from light water reactor coolant activation during power operation. For "quick and dirty" estimates, three inches of ordinary concrete reduces the dose rate of 1 Mev gammas by a factor of two.

10.   Typical shielding thickness for a pressurized water reactor plant are: (unless otherwise specified, ordinary concrete is used)

reactor primary shield	6'6"
secondary shield around coolant loop	3'
concrete containment wall	3'6"
concrete containment dome	3'
refueling canel wall	6'
refueling water depth above active moving assembly	10'

spent resin tank	4'
purification demineralizers	4'6"
low pressure injection pumps	2'6"
coolant breed holdup tanks	4'
waste drumming area	3'
spent fuel cooling demineralizer	3'
spent fuel coolant filter	2'6"
spent fuel coolant pumps and heat exchangers	0
labyrinth endwalls and valve gallery walls	1' to 1'6"

11. An useful approximation of fission-product source strength from typical power reactor is

$$S = \frac{100E}{T}$$

where S = source strength of fission product, Ci

E = thermal energy produced through fission, Mwd

T = time after reactor shutdown, years

(*Note:* The fission of 1 gram of uranium produces about 1 Mwd of thermal energy.)

SOLVED PROBLEMS

10-1 Check equation 10-2 against Figure 10.3 for 0.1 Mev gammas in lead.
Answer: For lead ($Z = 82$) and 0.1 Mev gammas, equation 10-2 gives

$$\sigma_{ph} = \frac{10^{-8}(82)^{4.05}}{(0.1)^{7/2}} = 1.8 \times 10^3 \text{ barns/atom}$$

$$= 1.8 \times 10^3 \frac{\text{barns}}{\text{atoms}} \times \frac{0.602 \times 10^{24} \text{ atoms}}{207 \text{ gram of lead}} = 5.23 \text{ cm}^2/\text{g of lead.}$$

This checks well against Figure 10.3.

10-2 Estimate the slab thicknesses of water and lead required to reduce 0.1 Mev gammas, fission neutrons, and thermal neutrons by a factor of 10^{10}.
Answer: Neglecting inverse-square spreading and buildup factors,

$$\frac{I}{I_0} = 10^{-10} = e^{-23} \quad \text{and} \quad e^{-23} = e^{-\mu x}$$

where for gammas in lead (Figure 10.3),

$$\mu = 5 \text{ cm}^2/\text{g} = 5 \times 11.35 \text{ cm}^{-1} = 56.75 \text{ cm}^{-1}$$
$$x = 23/56.75 = 0.406 \text{ cm} = 0.16 \text{ inch.}$$

For gammas in water (Figure 9.1),

$$\mu = 0.17 \text{ cm}^2/\text{g} = 0.17 \times 1 \text{ cm}^{-1} = 0.17 \text{ cm}^{-1}$$
$$x = 23/0.17 = 135 \text{ cm} = 4.44 \text{ feet}$$

For fission neutrons in lead (Table 10.2),

$$\mu = 0.12 \text{ cm}^{-1}$$
$$x = 23/0.12 = 192 \text{ cm} = 6.3 \text{ feet}$$

For fission neutrons in water (Table 10.2),

$$\mu = 0.0978 \text{ cm}^{-1}$$
$$x = 23/0.0978 = 236 \text{ cm} = 7.75 \text{ feet}$$

For thermal neutrons in lead,

$$\mu = \frac{1}{\sqrt{6}\,L} \qquad \text{(Problem 5-a)}$$

$$= \frac{1}{\sqrt{6}} \sqrt{\frac{\Sigma_a}{D}} \approx \frac{\sqrt{\Sigma_a}}{\sqrt{6}} \sqrt{3\Sigma_s} \qquad \text{(Equation 5-10)}$$

$$= \sqrt{\frac{0.006 \times 0.363}{2}} = 0.033 \text{ cm}^{-1} \qquad \text{(Table 3.1)}$$

$x = 23/0.033 = 697 \text{ cm} = 22.8 \text{ feet}$

For thermal neutrons in water (Problem 5-a and Table 5.1),

$$\mu = \frac{1}{r_t} = \frac{1}{\sqrt{6}\,L} = \frac{1}{\sqrt{6}\,(2.76)} = 0.148 \text{ cm}^{-1}$$

$x = 23/0.148 = 155 \text{ cm} = 5.1 \text{ feet}$

10-3 A 100 curies K-42 source, a 100 gram aluminum block, and a thick shield are arranged as shown. What is the dose rate at P due to once scattered photons?

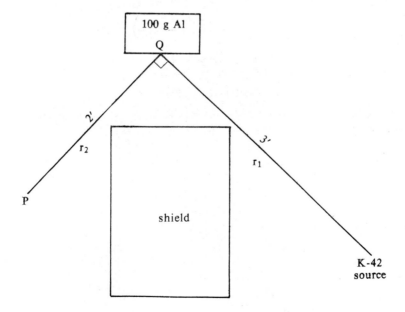

Answer: K-42 source emits 1.51 Mev gammas in 20% of the disintegrations. The source strength is therefore

$$S = 100 \times 3.7 \times 10^{10} \times 0.20 = 7.4 \times 10^{11} \text{ gammas/sec}$$

The flux at point Q is

$$\phi(Q) = \frac{S}{4\pi r_1{}^2} = 7.04 \times 10^6 \text{ gammas/cm}^2\text{-sec}$$

Since the scattering angle is $\pi/2$, equation 10-5 gives, for $a = 2.96$,

$$\frac{d\sigma(E, \theta)}{d\theta} = 2\pi \frac{r_0{}^2}{2} \frac{1}{(1 + a)^2} \left[1 + \frac{a^2}{1 + a} \right] = 0.0512 \text{ barn/electron-radian}$$

The aluminum block has been assumed to be a point (no absorption). The number of electrons in the block is

$$n = MA_0Z/A = 100 \times 0.602 \times 10^{24} \times 13/27 = 29 \times 10^{24} \text{ electrons}$$

The scattering rate at the block that results in photons going toward point P is

$$R = \phi(Q) \, (d\sigma/d\theta) \, n = 7.04 \times 10^6 \times 0.0512 \times 29 = 1.045 \times 10^7 \text{ gammas/sec-radian}$$

The differential area element per radian at a distance r_2 away from Q and orthogonal to r_1 is $(2\pi r_2)(r_2 \, d\theta)/d\theta$ and the flux at point P is thus

$$\phi(P) = \frac{1.045 \times 10^7}{4\pi r_2{}^2} = 448 \text{ gammas/cm}^2\text{-sec}$$

Now, equation 10-4 gives

$$\frac{1}{E'} = \frac{1}{1.51} + \frac{1}{0.51}\left(1 - \cos\frac{\pi}{2}\right)$$

or, $E' = 0.38$ Mev. Figure 9.2 gives 1 mr/hr $= 1250$ photons/cm^2-sec. Thus, the expected dose rate from the scattering of the aluminum block is

$$D = 448/1250 = 0.36 \text{ mr/hr}$$

10-4   Calculate the dose BUF for 2 Mev gammas in 20 cm and 200 cm water.
Answer: For water, $\mu_m = 0.05$ cm^2/g (Figure 9.1), and $\mu_0 = 0.05$ cm^{-1}. For $x = 20$ cm, $\mu_0 x = 1$ and Table 10.1 gives

$$B_r = 6.4 \, e^{0.076} + (1 - 6.4) \, e^{-0.092} = 2.0$$

For $x = 200$ cm, $\mu_0 x = 10$ and

$$B_r = 6.4 \, e^{0.76} + (1 - 6.4) \, e^{-0.92} = 11.55$$

10-5   A point P is located at the surface of an infinitely thick slab of uniform source strength S_V. Calculate the equivalent surface source density S_A.
Answer: Since P is at the surface, $b = 0$. Also, $b' = b + \mu_s(\infty) = \infty$. Using the formula for semi-infinite slab,

$$\phi(P) = \frac{S_V}{2\mu_s} [E_2(0) - E_2(\infty)] = \frac{S_V}{2\mu_s} (1 - 0) = \frac{S_V}{2\mu_s}$$

The equivalent surface source density S_A is evidently given by $S_A = 2\phi(P)$. The factor 2 accounts for the fact that half of the radiation goes in the opposite direction. Thus,

$$S_A = S_V/\mu_s$$

10-6 A point P is located with respect to two point gamma sources S_1 and S_2 as shown. The two sources are shielded by two solid spheres as indicated. How can the radii r_1 and r_2 be calculated if the total weight of the shielding materials are to be kept at a minimum and if the uncollided flux at P is given?

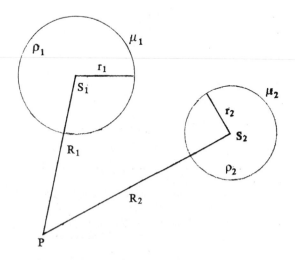

Answer: The given total flux at P is $\phi = \phi_1 + \phi_2$, or

$$\phi = \frac{S_1\, e^{-\mu_1 r_1}}{4\pi R_1^2} + \frac{S_2\, e^{-\mu_2 r_2}}{4\pi R_2^2} \tag{A}$$

If the weight of the two shields are $W_1 = (4/3)\pi r_1^3 \rho_1$ and $W_2 = (4/3)\pi r_2^3 \rho_2$ and if the total weight is to be a minimum, one must have

$$\frac{\partial \phi_1}{\partial W_1} = \frac{\partial \phi_2}{\partial W_2}$$

or,

$$\frac{\partial\left(S_1 \dfrac{e^{-\mu_1 r_1}}{4\pi R_1^2}\right)}{\partial\left(\dfrac{4}{3}\pi r_1^3 \rho_1\right)} = \frac{\partial\left(S_2 \dfrac{e^{-\mu_2 r_2}}{4\pi R_2^2}\right)}{\partial\left(\dfrac{4}{3}\pi r_2^3 \rho_2\right)}$$

i.e.,

$$\frac{S_1\mu_1}{R_1^2\rho_1}\frac{e^{-\mu_1 r_1}}{r_1^2} = \frac{S_2\mu_2}{R_2^2\rho_2}\frac{e^{-\mu_2 r_2}}{r_2^2}$$

or,

$$\frac{\mu_1}{r_1{}^2 \rho_1} \phi_1 = \frac{\mu_2}{r_2{}^2 \rho_2} \phi_2 \tag{B}$$

The values of r_1 and r_2 may then be obtained by solving equations A and B above.

10-7   What are the approximate fast neutron and gamma sources of a 100 kwt swimming pool research reactor?
Answer: Each fission produce 2.47 fast neutrons. Thus,

$$S_n = 100 \times 10^3 \text{ watts} \times 3.3 \times 10^{10} \text{ fission/sec-watt} \times 2.47 \text{ fast } n/\text{fission}$$
$$= 8.15 \times 10^{15} \text{ fast neutrons/sec.}$$

Since fission gammas account for about 20 Mev/fission, if each gamma is assumed to have 4 Mev, then

$$S_g = 10^5 \text{ watts} \times 3.3 \times 10^{10} \text{ fiss/sec-watt} \times 20 \text{ Mev/fiss} \times \tfrac{1}{4} \text{ gamma/Mev}$$
$$= 1.65 \times 10^6 \text{ gammas/sec.}$$

It must be noted that some of the neutrons will have to be retained inside the core for a chain reaction.

10-8   The Bulk Shielding Reactor (BSR) at Oak Ridge is a swimming pool reactor with core size of $15''W \times 15''L \times 24''H$. The core has volume fraction of water: 0.058; aluminum: 0.415; uranium: 0.002.
Estimate the depth of the water required to shield the reactor at 100 kw.
Answer: The desired attenuated levels are:

     fast neutron number flux $= 10 \ n/\text{cm}^2\text{-sec}$ (2 mrem/hr)
     gamma energy flux $= 1600 \text{ Mev/cm}^2\text{-sec}$ (2 mrem/hr)

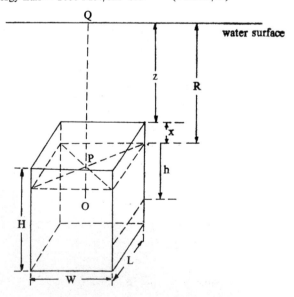

The first step is to transform the cubical core into a point source and the object is to locate the point P such that a point source

$$S_P = S_V HWL$$

located at P would be equivalent to the cubic core.

Let the absorption or removal properties of the core be represented by μ_c and that of the water by μ_w. The dose at point Q due to a point source at P is

$$D = S_P B(R) \frac{e^{-(\mu_c x + \mu_w z)}}{4\pi R^2}$$

The dose at point Q due to the volume source is

$$D = \int_{-x}^{H-x} S_V WLB(R + h) \frac{e^{-[\mu_c(x+h)+\mu_w z]}}{4\pi(R + h)^2}\, dh$$

where $B(R)$ and $B(R + h)$ are buildup factors and h is a dummy variable. Noting that

$$\frac{B(R + h)}{(R + h)^2} \approx \frac{B(R)}{R^2} = \text{constant}$$

and equating the two doses, one has

$$H = \int_{-x}^{H-x} e^{-\mu_c h}\, dh$$

or,

$$H = \frac{e^{\mu_c x}}{\mu_c} [1 - e^{-\mu_c H}] \tag{A}$$

This expression is true for both gammas and fast neutrons, although $x_g \neq x_n$, since $\mu_{cg} \neq \mu_{cn}$.

Using the data in Table 10.2 and the appropriate volume fractions, one has

$$\mu_{cn} = 0.583(0.0978) + 0.415(0.079) + 0.002(0.17) = 0.090 \text{ cm}^{-1}$$

Similarly, for 4 Mev gammas (see Problem 10-7),

$$\mu_{cg} = 0.583(0.034)(1) + 0.415(0.031)(2.7) + 0.002(0.044)(18.9)$$
$$= 0.056 \text{ cm}^{-1}$$

Substituting μ_{cn} and μ_{cg} into (A) separately, one has, for $H = 61$ cm,

$$x_n = 18.9 \text{ cm}$$
$$x_g = 22.5 \text{ cm}$$

In other words, the equivalent fast neutron point source and the equivalent gamma point source are not at the same location.

To calculate the depth of water z_n to shield the fast neutrons, one has

$$\phi_n = S_n \frac{e^{-(\mu_{cn} x_n + \mu_{wn} z_n)}}{4\pi(z_n + x_n)^2} \tag{B}$$

where $S_n = 8.15 \times 10^{15}$ (Problem 10-7), $\phi_n = 10$, and $\mu_{wn} = 0.0978$.

Solving equation (B) by graphical method, one has

$$z_n \approx 200 \text{ cm} \approx 7 \text{ feet}$$

for fast neutrons. No buildup factor has been used since removal cross-sections are used.

To calculate the depth of water z_g for gamma attenuation, note that

$$\phi_g = S_g B \frac{e^{-(\mu_{cg}x_g + \mu_{wg}z_g)}}{4\pi(z_g + x_g)^2} \tag{C}$$

where $S_g = 1.65 \times 10^{16}$ gammas/sec $= 6.6 \times 10^{16}$ Mev/sec (Problem 10-7),

$$\phi_g = 1600 \text{ Mev/cm}^2\text{-sec}$$

and the coefficients for the dose buildup factor B in water are

$$A = 4.5, a = -0.056, \text{ and } b = 0.117$$

Therefore,

$$B = 4.5 \, e^{0.056(0.056 \times 22.5 + 0.034 z_g)} - 3.5 \, e^{-0.117(0.056 \times 22.5 + 0.034 x_g)}$$

Substituting these values into equation (C) and solving the latter by graphical or trial-and-error method, one has

$$z_g \approx 510 \text{ cm} \approx 17 \text{ feet}$$

The actual depth of water for the BSR is 20 feet.

SUPPLEMENTARY PROBLEMS

10-a Obtain an expression for the uncollided flux due to a cylindrical surface source. The field point is in the radial direction, symmetrical with respect to the center of the cylinder.

10-b Using Table 10.1 and Figure 10.3, calculate the dose BUF for 1 Mev gamma in 1 cm and 20 cm thick lead. If the BUF are then written in the form: BUF $= 1 + c\mu t$, calculate the values of c.

10-c What are the limits of equations 10-5 and 10-7 for small and large photoelectron energies?

10-d A 500 Mwt PWR similar to the one described in problem 8-3 has a cylindrical core 7.5' tall and 3.2' in radius. Suppose that it is shielded by the following materials, in that order:

Material	Other function	Thickness
water	reflector and coolant	21 cm
steel	core barrel	2.5 cm
water	coolant	5 cm
steel	thermal shield	7.5 cm
water	coolant	5 cm
steel	pressure vessel	20 cm
water	—	90 cm
concrete	—	150 cm

What is the approximate fast neutron flux and gamma flux outside of each layer at the midplane of the reactor and along the radial direction? State your assumptions (or explain your model) clearly.

Chapter 11

FUEL CYCLE AND ECONOMICS

SINCE nuclear fuel is not completely expended during its first "lifetime" in a reactor, the reuse of the fuel creates the fuel recycle. In this chapter, an attempt is made to follow the uranium itself from the ore to the core and through reprocessing. This is summarized in Figure 11.1. Each of the major steps will be broken down and examined, with the emphasis being on light-water-cooled reactors (LWR). Nuclear fuel management and the economic aspect of the fuel cycle will then be presented briefly.

A. Mining

It is estimated that the uranium content in the earth crust is about four parts per million (ppm). It is as plentiful as lead and more plentiful than silver or mercury, but only 0.72% of the uranium is U-235. Promoted by the USAEC, exploration for uranium in the U.S. started in earnest in the early 1950's. Modern prospecting is done using airplanes with sophisticated electronic devices. Today, the richest uranum mines in the U.S. are located in the Colorado plateau region. The underground or open pit methods of mining uranium are much like that of other grade ore mines. The average uranium mine in the U.S. has an uranium assay of about 0.25% (*i.e.*, each ton or 2000 pounds of ore contains about 5 pounds of natural uranium), although other uranium sources available to the Western countries, such as those in Katanga and Canada may have uranium assay as high as 4%. Other minerals usually come with the uranium ore.

B. Ore Concentration

The uranium mills are located near the mines. These mills are designed to purify and concentrate thousands of tons of ore per month into natural uranium oxides, the so-called "yellow cake", which contains 70% to 90% of U_3O_8. Forecast requirements for the period 1973–82 amount to some 235,000 tons of uranium oxide for the U.S. alone. The major steps in ore concentration are:

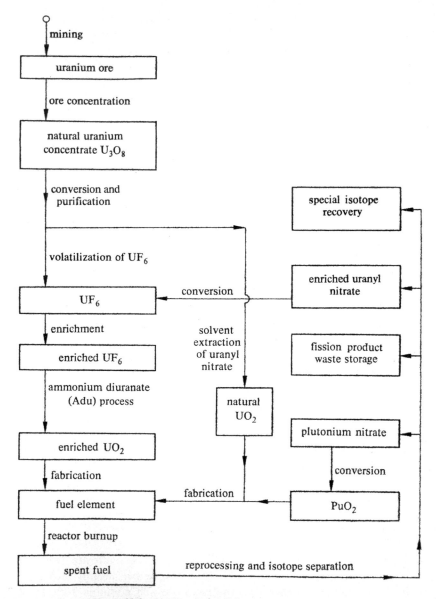

Figure 11.1 Fuel cycle for light-water-cooled reactors.

1. The ore is transported by trucks or trains to the mill.
2. The ore is crushed and milled to a sugar-like consistency. Standard metallurgical procedures such as screening, washing, flotation, and gravity separation follow. Water is added to create a slurry.
3. The slurry is pumped to large leaching tanks where sulfuric acid (or sodium carbonate) is added to dissolve all the uranium.
4. The sand or tailings is separated by centrifugal or rake-type classifiers and disposed of, leaving the solution that contains uranium.
5. The uranium compound is now ready to be extracted from the acid solution by the methods of solvent extraction or ion-exchange.

In the more popular solvent extraction method, the acid solution flows through an acid-strip to the precipitation circuit. When heat, ammonia, and air are added, the uranium becomes insoluble and is settled out in large thickener tanks. After filtering and drying, the final product, "yellowcake", as it is often called, results.

In the ion-exchange method, the dilute solution is passed through a fixed bed of ion-exchange resin (which consists of a loosely cross-linked polymerized organic structure with a number of active groups attached). The basic process is one of unsteady-state diffusion in a liquid-solid system. Again, the final powdered product contains a high percentage of U_3O_8.

Yellowcake is a crude oxide or salt concentrate that contains a high percentage of U_3O_8. The oxide is sold at a price of US\$6.25 to US\$8 per pound. A mill is designed to produce about 1000 tons of U_3O_8 per year.

C. Concentrate Purification

The next step in the uranium fuel production is purification. The end products are either pure uranium hexafluoride ("hex"), or pure uranium dioxide ("brown oxide"), both unenriched.

To obtain pure uranium dioxide, UO_2, from uranium ore concentrate and impurities, the method of solvent extraction is used.* The concentrate is first dissolved in nitric acid and the resulting uranyl nitrate and impurities are fed to the top of an extraction column. An organic solvent, tributyl

---

* The solvent extraction method is a diffusion operation and is used to separate the constituents in a solution, either aqueous or organic. An organic solvent, which is immiscible with water, is used. The fact that the solubilities of the constituents are different in water and in the organic solvent makes it possible to achieve partial separation of the constituents when organic liquid is bought into contact with the aqueous solution.

phosphate (TBP) diluted with kerosene (an inert hydrocarbon), is forced to flow upward through this column and the uranyl nitrate is extracted into this organic medium and collected at the top of the column. The organic solution is now forced to flow upward through a stripping column in which water is used to back-extract the uranium into the aqueous phase. Upon controlled evaporation, an essentially impurity free product, called uranyl nitrate hexahydrate, $UO_2(NO_3)_2 \cdot 6H_2O$, or simply UNH, results. When heated at about 1000°F and excess water removed, the nitrate decomposed and one has uranium trioxide ("orange oxide"), UO_3. Pure unenriched UO_2 can then be obtained by reducing UO_3 with hydrogen at about 1100°F in a fluidized-bed reactor. The hydrogen is supplied by thermal dissociation of ammonia.

To obtain pure uranium hexafluoride, UF_6, from uranium ore concentrate, the method of fluorination and volatilization is used. The U_3O_8 powder is first chemically reduced to UO_2, similar to the way UO_3 is reduced, but at slightly higher temperature. Two other chemical reactions are then designed to take place:

$$UO_2 + 4HF \xrightarrow[900°F]{1200°F} 2H_2O + UF_4 \text{ ("green salt" powder)}$$
$$UF_4 + F_2 \longrightarrow UF_6$$

The UF_6 is a solid at STP but sublimes to gas at 133.5°F and 1 atm. The UF_6 gas from the last reaction can be collected in a cold trap and can be purified further by the technique of fractional distillation* at moderate temperatures and pressures. A fluorination plant is designed to have a conversion capacity of about 10,000 tons per year.

D. Enrichment

Today, all uranium enrichment in the U.S. is government operated. Beginning January 1, 1971, the only means of getting enriched uranium in the U.S. is to purchase it or pay for the toll enrichment by the Government. Beginning July 1, 1973, all private operators who have been loaned enriched uranium by the Government, must purchase the title from the Government. At present, pure UF_6 is the only feed material for the U-235 enrichment process. The USAEC has three plants. In 1969 alone, nearly 700,000 kg of enriched UF_6 were shipped out to U.S. processors and fabricators. There are three important methods for uranium enrichment:

---

* Fractional distillation is based on the fact that various materials vaporize at different temperatures and pressures.

1. Gaseous Diffusion

Since all molecules at a given temperature have the same average energy, the heavier ones necessarily travel slower. When natural UF_6 gas is forced to diffuse through a porous barrier* by a small pressure difference at the two sides of the barrier, the slightly faster $U^{235}F_6$ molecules have a slightly better chance of getting through than the heavier and slower $U^{238}F_6$ molecules, since the frequency of collisions with the barrier is higher for the former. The UF_6 on the lower pressure side of the barrier is therefore slightly enriched. If many stages are connected in series, higher and higher enrichment can be obtained. Theoretically, it takes about 500 stages to obtain 3% enrichment and about 1500 stages to obtain 40% enrichment. In practice, about 1500 stages are required to produce 3% enrichment, the kind used in light-water-cooled power reactors. There is, however, no need to have parallel connections, since each stage can be made quite large. The cost per gram of 40% enriched uranium is about 3 times that of natural uranium.

In 1962, when the three U.S. gaseous diffusion plants were operated at over 90% capacities, the amount of electricity used in the plants to operate the pumps and associated equipments was about 4.7×10^{10} kwh and was about 5% of the total electricity generated in the country that year. If pushed at full capacities, these three plants can produce 5000 tons of 2.6% enriched uranium annually, enough for 200 light-water reactors of 650 Mwe each to operate continuously.

2. Gas Centrifuge Process

When a gas such as UF_6 is spinned, the heavier $U^{238}F_6$ molecules experienced a larger centrifugal force than the lighter $U^{235}F_6$ molecules and the former move outward more, relative to the latter. The design of the apparatus is basically that of a high speed rotor. The inherent problems are friction and wear, oscillation, and economy. Since most of the information is still classified and no major pilot plant has been built, it is difficult to evaluate accurately the potential of this technique. However, the following general conclusions are of interest:

a   The centrifuge process is very applicable for small separation plants. Only 10 to 20 stages in series are required to produce 3% enrichment material, although parallel systems are necessary for large volume works.

---

* The nature of the barrier (or membrane) is still classified. It must be acknowledged. however, that it is a very ingenious feat to develop a barrier that has pore diameters of about 5×10^{-5} mm, can withstand the pressure difference, will resist the highly corrosive UF_6 gas, and does not get clogged up by the solid decomposition products of UF_6. It is reported that the "barrier" is flexible and that sintered nickel powder may have been used.

b)   Approximately 2 million centrifuges may be required to produce the same capacity as a single large diffusion plant. Each complete centrifuge apparatus may be about 5 feet tall and 1 foot in diameter. The internal rotor that contains the UF_6 gas may be 3 feet tall and 8″ in diameter. The centrifuge plant may use only 10% the power of an equivalent gas diffusion plant.
c)   Technologically speaking, it is probably the easiest method in producing weapons-grade materials.
d)   Since parallel systems can be easily converted to series system, higher enrichment may be easily obtained. For this reason, a centrifuge process plant may not be easily licensed under a nonproliferation treaty.

3.   *Separation Nozzle Method*

Figure 11.2 shows the basic operation principle of this recently developed technique. The research is developed in Germany, directed by professor

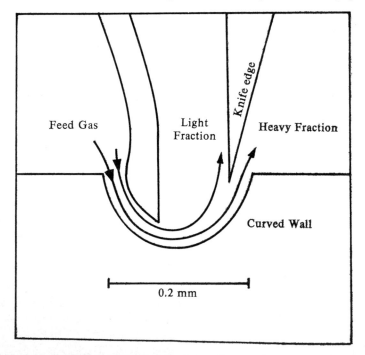

Figure 11.2   Section through a separation nozzle arrangement.

E. W. Becker. A 10-stage trial-scale pilot plant was first operated in late 1967 and further research and development is going on actively.

The idea is quite simple. The feed gas contains 5 mole % UF_6 gas and 95 mole % He gas. The helium is there to enhance the flow velocity and therefore isotope separation. The feed gas is under a pressure equivalent to 600 mm Hg. The spaces to the right of the nozzle have a lower pressure (by a factor of about $\frac{1}{4}$). The fixed, curved wall and the knife edge separate the feed gas into two portions. The light fraction contains enriched UF_6 and higher percentage of He than the feed gas while the heavy fraction is depleted in both He and U-235.

For the pilot plant, each stage contains 100 elements and each element contains 10 slit-shaped separation nozzle systems 2 meters long and arranged on the circumference of a 10 cm diameter aluminum tube. Preliminary results indicate that about 500 trial-scale cascade stages will give a 3% enrichment. For industrial operation, no parallel systems are necessary.

At present, it appears that the following factors are of interest:

a) A cascade chain such as that described above has a capacity of roughly one-third that of the Oak Ridge gaseous diffusion plant.

b) The investment cost of the nozzle separation process plant would be lower than that of a gaseous diffusion plant. The fact that no porous barrier is needed may reduce the maintenance cost considerably.

c) Although the specific energy consumption in operating a gaseous diffusion plant is high, the nozzle separation process requires still higher specific energy consumption (by a factor of 2.2 for the pilot plant, but this may be reduced considerably upon further research and development). Thus, according to the pilot plant, it takes 10% of the energy produced by the final fuel in a light-water reactor to enrich the fuel in the first place.

4. Others

There are other methods of separating uranium isotopes. One of these is the electromagnetic separation, based on the principle that ions of isotopes of different masses travel with different radii of curvatures in a magnetic field. This method was used during the Manhattan Project. It was found that the process is reasonably effective in getting material of high enrichment, starting with slightly enriched material. The technique is useful in scientific research but it is not good for industrial production of uranium isotopes.

E. Ammonium Diuranate (Adu) Process

Next, the enriched UF_6 is shipped from a fuel enrichment plant to a fuel assembly plant. The first step is to convert the UF_6 into UO_2 by the Adu

o

process: The UF_6 is hydrolyzed with a dilute solution of ammonia. The precipitate of ammonium diuranate is formed. After filtering and drying, this is heated at high temperature in a mixture of steam and hydrogen (the hydrogen being supplied by the thermal disassociation or "cracking" of ammonia) to obtain the uranium dioxide solid.

$$UF_6 \xrightarrow[\text{heat}]{\text{H}_2\text{O,NH}_4\text{OH}} (NH_4)_2U_2O_7$$

$$(NH_4)_2U_2O_7 \xrightarrow[\text{heat}]{\text{H}_2,\text{H}_2\text{O}} UO_2$$

F. Uranium Dioxide

The final pure product, UO_2, contains 88.15% uranium and is desired as power reactor fuel for several reasons. Basically, uranium dioxide is a crystal with a face-centered cubic (f.c.c.) structure and a theoretical density of 10.97 g/c.c. The melting point is over 2800°C (5100°F). The fracture strength is about 10^4 psi. The thermal expansion coefficient is about $10^{-5}/°C$ between 0°C and 1000°C. The thermal conductivity is 0.02 cal/sec-cm-°C at 20°C. It reacts with oxygen (or air) readily and changes to a higher oxide.

The pellets used in light-water-cooled power reactors are high-density ceramic, formed by pulverizing, pressing, sintering, and grinding of the oxide. The average density of the pellets is about 95% of the theoretical density of UO_2. As a ceramic, uranium dioxide has the general physical and chemical stability properties at high temperature and under irradiation. The fact that it is inert to attack by hot water, that it tends to retain a large portion of the fission gas, and that it has a high melting point make it very desirable as a safe fuel for light-water-cooled reactors. It is essential that the density of the ceramic be made as high as possible in order that the relatively low thermal conductivity be increased, the fission gas retention ability is enhanced, and the uranium atom density is increased. It should be mentioned that the thermal conductivity, for example, of the ceramics depend noticeably on the method by which the ceramics are formed.

A typical modern 800 Mwe BWR plant requires an initial reactor core loading of about 135 tons of UO_2 fuel. For a PWR with equivalent power output, about 100 tons may be required.

G. Fabrication

Zirconium alloys are used as fuel cladding because of their strength, corrosion resistance, and low neutron cross-sections. Due to the rigid requirements of nuclear design, the making of the zirconium alloy cladding tubes is quite complicated and expensive. A zircaloy fuel rod (such as that described in Problem 8-3) is made by filling the tube with high-density uranium dioxide ceramic pellets, evacuating, backfilling with helium, and sealing by welding zircaloy end plugs in each end. The cladding must be strong enough to withstand the pressure inside a reactor. There is a plenum space above the pellets to accommodate the excess fission gases and a spring in this space keeps the pellets in position during transit. The helium gas is there to improve the thermal conductivity between the pellets and the cladding.

H. Burnup

For a reactor to operate at a power level of 1 Mwt, the U-235 atoms fission at a rate of about 1 g/day. The actual U-235 consumption rate for a 1 Mwt reactor is higher and is about 1.24 g/day, because of non-fission capture. A widely used unit for measuring burnup is megawatt-days per metric ton of U-235 fuel, MWD/T. A metric ton, or tonne, is 10^6 gram or 2205 pounds. It is evident, therefore, that if all the U-235 is consumed, by fission or otherwise, the energy release would be about 800,000 MWD/T.

When a fission reaction occurs in a fuel element, the microscopic structure of the material is damaged near the fission site. The damage is called radiation damage and affects the properties of the fuel in terms of structural integrity, fission gas retention capability, etc. It is found that material failure occurs when there is more than an one percent burnup, *i.e.*, when more than 1% of the atoms (U-235 and others) have undergone fission or when the energy release is around 15,000 MWD/T, depending on the fuel element. For a 3% enriched fuel element, this amounts to 33% of the U-235 fissioned (not just neutron absorption). One should, however, keep in mind that the burnup is not uniform inside a reactor. The production of fission gases poisonous to neutrons is another factor in the determination of the amount of burnup allowed before a fuel assembly is removed from a reactor for reprocessing.

In practice, a 800 Mwe BWR with 135 tons of UO_2 fuel in the initial reactor core loading would have an annual discharge rate of about 22 tonnes of uranium for reprocessing. This is approximately one-fifth of the core loading. A 800 Mwe PWR with 100 tons of UO_2 fuel would have an annual discharge rate of about one-third of the core loading, *i.e.*, approximately 33.5 tonnes of uranium.

I. Reprocessing

At the time this text was written, there were only three commercial reprocessing plants in operation in the U.S. (although a few more are scheduled to be in operation soon). The plant uses the "chop-leach" method in conjunction with the so-called Purex solvent extraction process. It is designed to handle a maximum of about 450 metric tons of spent fuel annually.

The highly radioactive spent fuel assemblies of a reactor are first stored under water near the power plant site for a few months to allow the short-life radionuclides to decay away. The not-so-radioactive assemblies are then shipped, under heavy shield, to the reprocessing plant. They are then mechanically sheared (chopped) into small pieces and leached with nitric acid. (In another method, used by the AEC for years, the entire rod, including the cladding, is dissolved, creating more liquid waste). The radioactive nitric acid solution is then subjected to the Purex* solvent extraction method described earlier and the radioactive fission fragments are removed. The final products of the reprocessing plants are separated uranium nitrate solution, plutonium nitrate solution, and neptunium nitrate solution. The uranium is of lower enrichment than that of the original fuel. The uranium loss is at about 0.1 %.

The non-radioactive uranium and plutonium nitrate solutions may then be converted, enriched, and used as described earlier.

The plutonium cycle is at an early stage of development. One of the difficult but manageable facts is that plutonium is highly toxic and hazardous to handle. The final product, PuO_2, may be mixed with uranium dioxide as fuel. The first Pu recycle plant is scheduled for operation in 1977.

J. Waste Disposal

The highly radioactive aluminum nitrate and zirconium type waste have to be disposed of safely. The treatments of radioactive wastes are different for solid, liquid, gaseous, high level, and low level wastes. Studies are being made for the most effective (safe and economical) method of waste disposal. Generally speaking, the methods involve solidification, calcination, volume reduction, and ground or deep sea burying. The U.S. does not practice sea disposal. As far as safety is concerned, the major danger is the contamination of underground water and marine lives, as these may become the intake for the human race at a later date, directly or indirectly.

---

* **Purex** = *P*lutonium *u*ranium *ex*traction

K. Fuel Loading

Figures 11.3 and 11.4 show the initial and future loading of a typical **PWR** plant. According to the refueling mode, about a third of the fuel is removed from the core annually, fresh fuel is added ,and the fuel locations are changed. Such multi-region core management allows 50% greater burnup (megawatt days of heat/metric ton) than that of batch fuel loading.

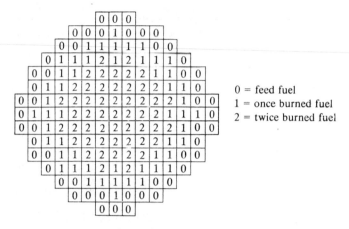

0 = feed fuel
1 = once burned fuel
2 = twice burned fuel

(a)   Initial loading pattern

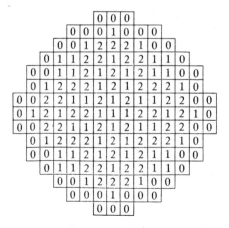

(b)   Future loading pattern

Figure 11.3   Westinghouse PWR fuel loading patterns.

L. Fuel Economics

In order that one can get a general representative idea of the approximate cost involved, data for a typical Westinghouse PWR is given in Tables 11.1 through 11.4.

Unlike conventional fuel, whose cost depends largely on the purchase price, the nuclear fuel economics depends strongly on the management variations. Time is therefore a most important factor. Generally speaking, the nuclear fuel cost can be separated into two groups, the depreciation charge and the carrying charge. The depreciation charge is the difference between initial and final investment. The carrying charge rate is assumed to be 12% annually, including interests and taxes. The fuel cost is summarized in Table 11.4.

TABLE 11.1   Fuel economic design parameters of typical PWR

reactor region power	1,000 Mwt
region loading	29,500 kg U
enrichment	3.3% wt. U-235
burnup	31,500 MWD/MTU
discharge loading	28,000 kg U
discharge enrichment	0.9% wt. U-235
discharge plutonium (fissile)	180 kg
capacity factor	85%
region lifetime	36 months
total energy produced	76.1×10^{12} BTU

TABLE 11.2   Fuel cost assumptions for typical PWR

yellowcake	\$7.50/lb. of U_3O_8
conversion to UF_6	\$2.52/kg of U in UF_6
enrichment*	\$26/separative work unit
	(0.2% tails concentration)
design and fabrication	\$80/kg of U loaded
ship spent fuel	\$5/kg of U discharged
reprocessing	\$37/kg of U discharged
plutonium value	\$8,000/kg of fissile Pu

* The separative work unit (s.w.u.) is the standard measure of enrichment service. Five s.w.u. are required per kg of U with 3.3% wt. U-235

TABLE 11.3 Fuel costs and investment timetable of a typical 1000 Mwe PWR

Fuel cycle step	Month	10^3 US$	
buy U_3O_8 as yellowcake	−13	3508	10,129*
purify and convert to UF_6	−10	451	
enrichment of UF_6	−7	3810	
design and fabricate fuel assemblies	−11 to −2	2360	
			10,129*
start operation	0	0	
end operation	36	0	
			1284†
ship spent fuel	43	140	
reprocess spent fuel	45	1036	
uranium credit	46	−1020	
plutonium credit	46	−1440	

* Total pre-operation investment in fuel
† Net residual value

TABLE 11.4 Fuel cost summary (¢/million BTU)

Component	Depreciation	+ carrying charge	= total
uranium and plutonium	6.98	3.75	10.73
design and fabrication	3.10	0.76	3.86
spent fuel shipment	0.19	−0.05	0.14
reprocessing	1.36	−0.37	0.99
total	11.63	4.09	15.72
			(¢/million BTU)

M. Nuclear Power Plant Cost

With the understanding that nuclear fuel is the major cost of a reactor plant in its designed life of about 40 years, it is appropriate to get an idea of roughly how much is the capital cost of a typical nuclear plant. Referring again to the Carolina Power and Light Company's Brunswick Plant I and II, described in Chapter 4, a breakdown of costs for the two 840 Mwe BWR by the General Electric Company is given in Table 11.5.

TABLE 11.5   Cost breakdown of two adjacent 840 Mwe BWR

	10^6 US$
land and land rights	1.5
structure and improvements	67.8
reactor plant equipment	121.7
turbine generator plant	71.3
accessory electrical equipment	15.9
miscellaneous power plant equipment	4.4
general plant (main transformer and switchyard)	7.6
interest during construction	40.8
fuel inventory	54.0
total estimated cost	385.0

Concerning the items listed in Table 11.5, it may be noted that the land and land rights cover the required adjacent exclusion area. When two or more plants are built on the same site, there is a definite saving. The reactor plant equipment covers the reactor equipment, heat transfer equipment, fuel handling and storage facilities, off site fuel reprocessing and refabrication, waste disposal, instrumentation and control, feedwater supply and treatment, and steam condenser and feedwater. The normally indirect construction cost of engineering design and inspection is also included.

Obviously, the capital and operating costs of a nuclear plant depend not only on the type, size, unit number, and location, but also on the construction time, management of fuel loading, and other economic factors such as interests, taxes, and inflation. Nuclear plant cost optimization is, indeed, an art and subject by itself.

N.   Uranium Fuel Reserves

According to a recent report (Wash-1942), U.S. annual uranium requirements would increase from 18,000 tons of U_3O_8 in 1975 to 120,000 tons in 1990. Current known reserves (productible at $80/pound or less) amount to 273,000 tons. An additional 450,000 tons of $8 uranium is estimated to be in known favorable geologic environments. With $15 uranium, the total resources must be expanded to support post-1990 requirements.

Chapter 12

PROGRESS IN REACTOR ENGINEERING

THE OBJECT of this chapter is to give a general assessment of the state of the nuclear reactor field and a projection for the year 1980. Some of the data are from the USAEC (See, for example, The Nuclear Industry, by AEC, December, 1969). It is hoped that the readers will be more aware of the scale and trend of the nuclear field, especially in the United States. In particular, the progress in the fields of high temperature gas cooled reactor and liquid metal fast breeder will be presented. A short discussion on the outlook for controlled fusion power will then conclude this chapter.

A. The Nuclear Industry

Table 12.1 shows some of the interesting statistics of the U.S. nuclear industry. The data are often rounded, since only the general magnitude is of interest here. Table 12.2 shows the USAEC budget figures for the fiscal years of 1971, 1972, and should reflect to a certain extent the emphasis and areas of operation of the AEC.

B. Contemporary Uses of Nuclear Power

Aside from the extensive classical uses of radioisotopes in medicine, earth satellites, industries, agriculture, scientific research, etc., in relation to the peaceful application of nuclear power, three topics are of particular contemporary interest. All three are related to nuclear reactors and are discussed briefly below.

1. *Space Systems*

On November 19, 1969, the Apollo 12 astronauts set up experiments on the moon's surface, powered by a compact plutonium radioisotope power system, the SNAP-27 (Systems for Nuclear Auxiliary Power-27). It contains 8.36 pounds of Pu-238. The generator is 16″ in diameter, 18″ in height, and can generate over 70 watts of electricity. While that was going on, the joint NASA-AEC nuclear rocket program was well on the way. An immediate

TABLE 12.1   Selected statistics of the U.S. nuclear industries

	1966	1967	1968	1969	1970	1980
Value of net orders received by industries for selected atomic energy products such as reactors and associated power plant components	0.71 B$	1.69 B$	1.02 B$			
Employment in government facilities		101,493	100,972			
Employment in private facilities		36,499	43,383			
Export of selected nuclear commodities such as reactors, fuels, and instrumentations	120 M$	97 M$	102 M$		59 M$	14 M$
Electric utility generating capability: nuclear					20 Bwe	150 Bwe
others					340 Bwe	550 Bwe
Announcements of new steam electric plant additions: nuclear	22 Bwe	26 Bwe	15 Bwe	8 Bwe		
others	20 Bwe	32 Bwe	24 Bwe	33 Bwe		
Cumulative investment of nuclear construction investment and fuel cycle cost (based on $200/kwe for construction cost)					1.5 B$	50 B$
Annual requirements of U_3O_8 for nuclear power: (in kilotons) U.S.A.					7.5	34
foreign					6.0	38
Cumulative requirements of U_3O_8 for nuclear power since 1970 (in kilotons): U.S.A.					7.5	208
foreign					6.0	134
Yellow cake resources (costing $10/lb or less to mine and process), in kilotons Reasonably assured: U.S.A.				300		
other non-communist countries				550		
Estimated additional: U.S.A.				370		
other non-communist countries				400		
Yellow cake resources (costing $10/lb to $30/lb to mine and process), in kilotons Reasonably assured: U.S.A.				350		
other non-communist countries				950		
Estimated additional: U.S.A.				650		
other non-communist countries				1350		
Uranium processing and fabrication market (fabrication cost based on $90/kg of contained uranium): initial fuel						

	103,000	160,000	112,000	243,000	189 M$	352 M$
replacement fuel						
Enriched uranium shipped abroad for foreign reactors, kgs U	103,000	160,000	112,000	243,000		
Plutonium recovered from domestic commercial power reactor fuels, kgs					270	20,000
Annual requirements for zirconium in U.S. reactors:						
million feet of tubing					4.5	28.4
million pounds					2	6.7
Reactor pressure vessels market (based on 60% PWR, 40% BWR)					12 M$	140 M$
Steam generators market					14 M$	160 M$
Pressurizers market					1 M$	11 M$
Reactor feed pump market					5 M$	57 M$
Valves, pipes, and tanks market					14 M$	160 M$
Instrumentation and controls, control rods, and control rod drive mechanism market					20 M$	230 M$
Turbine generators market					80 M$	880 M$
Turbo-generator equipment (condensers, cooling towers, heaters, etc.) market					10 M$	110 M$
Containment structures market					14 M$	160 M$
Total number of domestic civilian power and research reactor projects under contract, design, or construction				83		
Total estimated cost of the projects above				11 B$		
Irradiated fuel reprocessing market (@ $33,000/MTU)					1.8 M$	97 M$
Low level radioactive waste management market					1.5 M$	10 M$
Reactor resin waste handling and burial market (based on 500 cubic feet of resin discharged per year per 100 Mwe at 1980 with 150 Bwe)					250 k$	3.6 M$
Annual spent fuel processed, in tons					70	3600
Volume of high level waste, if solidified (based on 1 cubic foot of solidified waste for 100 gal of liquid waste per 10 Bwd thermal, with volume reduction factor of 13.4) in cubic feet					170	9700
Domestic shipment (in kilotons) of: U_3O_8					7.5	34
normal UF_6					9	43
enriched UF_6					1.8	9
UO_2					1.4	6.6
new fuel elements					1.8	7.6
irradiated fuel					0.07	3.8

TABLE 12.2   The AEC budget for fiscal year 1973

Summary of Appropriations
(in thousands of U.S. dollars)

	FY 1971	FY 1972	FY 1973
Appropriations:			
Operating Expenses	$1,929,160	$1,963,430[a]	$2,122,830
Plant and Capital Equipment:			
Capital Equipment Not Related to Construction	173,050	153,296	167,080
Construction	206,050	199,954	273,230
Total Plant and Capital Equipment	379,100	353,250[b]	440,310
Total Appropriations (NOA)	2,308,260	2,316,680	2,563,140
Expenditures (Outlays):			
Operating Expenses	1,881,491	1,962,406[c]	2,036,400
Plant and Capital Equipment:			
Capital Equipment Not Related to Construction	179,235	158,833	148,826
Construction	242,889	236,582	236,874
Total Plant and Capital Equipment	422,124	395,415[d]	385,700
Total Expenditures	$2,303,615	$2,357,822	$2,422,100

[a] Includes $13.3 million for proposed FY 1972 supplemental appropriation.
[b] Includes $9.0 million for proposed FY 1972 supplemental appropriation.
[c] Includes $10.0 million for proposed FY 1972 supplemental appropriation.
[d] Includes $1.4 million for proposed FY 1972 supplemental appropriation.

Operating Expenses
(in thousands of U.S. dollars)

	FY 1971 Actual	FY 1972 Estimates	FY 1973 Estimate to Congress
Accrued Costs:			
Nuclear materials	$ 362,242	$ 386,502	$ 426,600
Weapons	827,966	845,346	878,500
Reactor development	428,375	451,653	485,200
Physical research	270,765	267,800	282,800
Biology and medicine	88,117	90,845	94,500
Training, education and information	12,859	12,293	12,400
Isotopes development	6,464	5,900	5,900
Civilian applications of nuclear explosives	7,355	6,900	6,800
Community	7,805	5,080	5,000
Regulation	15,684	22,247	26,500
Program direction and administration	116,751	120,788	120,500
Security investigations	6,096	7,328	7,300
Cost of work for others	35,697	17,559	14,000
Adjustment to prior year costs	1,088	—	—
Total Program Costs, Funded	2,187,264	2,240,241	2,366,000

TABLE 12.2 (cont.)

Change in Selected resoirces	—7,400	81,882	92,300
Gross Obligations	2,179,864	2,322,123	2,458,300
Revenues applied	—329,336	—258,676	—317,000
Net Obligations	1,850,528	2,063,447	2,141,300
Unobligated balance, start of year	—41,017	—119,507	—18,470
Unobligated balance, end of year	119,507	18,470	—
Appropriation Transfers	142	1,020	—
Appropriation (NOA)	$1,929,160	$1,963,430[a]	$2,122,830

[a] Includes $13.3 million for proposed FY 1972 supplemental appropriation.

major objective is the development of the NERVA (Nuclear Engine for Rocket Vehicle Application) nuclear reactor rocket engine, which started in 1960. It will be 24' long, have about 75 kilopounds of thrust, 825 seconds of specific impulse, about an hour of endurance capability, and an extremely high reliability of multiple restarts in space. The reactor generates upwards of 5 Bwt. The ground-experimental "hot" engine (XE) test program was successfully completed in August 28, 1969. The NERVA engine achieves its propulsion capability by heating liquid hydrogen from $-420°F$ to over $+4,000°F$ in a compact reactor power plant. The hydrogen's tremendous expansion through the rocket nozzle produces the propulsive thrust. One promising mission is the manned Mars landing in the 1980's. The payload leaving earth orbit for Mars would be about 500,000 kg and would be able to support six to eight men during the 450-day roundtrip.

2. *Plowshare—Peaceful Uses of Nuclear Explosives*

The successful use of underground nuclear explosive to stimulate natural gas fields of low productivity has been demonstrated in the last few years. A study of using 2 to 3 kiloton underground nuclear explosive for excavation purpose is being actively studied by the Atlantic-Pacific Interoceanic Canal Study Commission. The State of Arizona is investigating the possibility of water management with nuclear explosive and there are many other possible industrial uses of nuclear explosives. For excavation application, the AEC projects a charge of $350,000 for a 10 kiloton yield and $600,000 for a 2 megaton yield. The charges cover nuclear materials, fabrication and assembly, and arming and firing services. As an illustration, an estimate was made back in 1964 on a new canal at the Panama. The route would require a cut of about 48 miles long with depths up to 1100 feet. The maximum yield for explosives is some 10-megaton and the total yield is around 170 megatons. The total cost is estimated to be $650-million, including $30-million for site survey and design, $290-million for nuclear excavation, $220-million for general construction, $30-million for engineering, and $80-million for contingency. It would cost $2-billion for conventional method.

3. *Pollution*

At the turn of this decade, one of the major issues and problems of public concern is pollution. There are three types of pollution: Chemical and waste pollution, radioactive substance pollution, and thermal pollution, all basically due to the population increase, growth of industries, improved standard of living, and power demand. (Another environmental problem is that of excessive noise, which may be considered as acoustic pollution.)

The chemical and waste pollution includes the industrial and domestic waste and garbage such as chemicals dispersed into the atmosphere, rivers, lakes, and sewage system, automobile exhaust fume, sulphur dioxide released by fossil steam plants, off-shore oil spill, garbage, etc. The nuclear engineers are not under fire as far as this type of pollution to the environment is concerned. Table 12.3 shows the air pollution emission in the United States.

TABLE 12.3   Air pollution emission—United States*

| Pollutant | Million tons per year | % of total | Sources (Millions of tons per year) | | | | |
			Power plants†	Industries	Auto-mobiles	Space heating	Refuse disposal
carbon monoxide	72	50.7	1	2	66	2	1
sulphur oxides	26	18.3	12	9	1	3	1
nitrogen oxides	13	9.2	3	2	6	1	1
hydrocarbons	19	13.4	1	4	12	1	1
particulates	12	8.4	3	6	1	1	1
total	142	100	20	23	86	8	5
% of total			14.1	16.2	60.6	5.6	3.5

* From "The Sources of Air Pollution and Their Control", U.S. Dept. of Health, Education, and Welfare, Washington, 1967.)

† Author's note: Mainly from fossil plants.

The radioactive substance pollution may originate from wastes from nuclear ore processing, nuclear fuel reprocessing, contaminated materials, and by-products related to radioactive material treatment and production. Surveillance programs have shown that radioactive substances (such as tritium and Krypton-85) released by normal nuclear power plant operation is so negligibly low that detecting them is difficult. By far the most important problem is that of radioactive waste treatment. When sea disposal is used, a small amount of the long half-lived substance could conceivably eventually get back to the human body through aquatic creatures. When land burial is practiced, it is essential to know such factors as underground water flow, earthquake, etc. in order that drinking water may not ever be affected. At present, all phases of the radioactive substance pollution are well under control, with the possible exception of radioactive wastes from nuclear ore processing.

Thermal pollution is the subject of much arguments. It is the simple law of thermodynamics that a power production plant must have heat waste. As a society demands more and more power, the power plants get larger and more numerous. The increasing heat waste may be disposed of into the air through cooling towers and into the lakes, rivers, and seas. The question is how much the water temperature rise is affecting the aquatic life and wild-life cycle and what should be done about it, in view of the fact that more and more power will be required in the future. The following facts are of interest in the understanding of the thermal pollution problem:

a)   As Figure 12.1 shows, nuclear stations are more sensitive to condenser temperature and, as far as heat production rate is concerned, it makes good engineering sense to push the heat rejection temperature higher. Generally speaking, nuclear power requires 40% more cooling water than fossil fuel generation and rejects 50% more waste heat.

b)   The Water Quality Act of 1965 states, in part, that

During any month of the year, heat should not be added to a stream in excess of the amount that will raise the temperature of the water (at the expected minimum daily flow for that month) more than 5°F. In lakes, the temperature of the epilimnion in those areas where important organisms are most likely to be adversely affected should not be raised more than 3°F above that which existed before the addition of heat of artificial origin. The increases should be based on the monthly average of the maximum daily temperature. Unless a special study shows that a discharge of a heated effluent into the hypolimnion will be desirable, such practice is not recommended and water for cooling should not be pumped from the hypolimnion to be discharged to the same body of water.

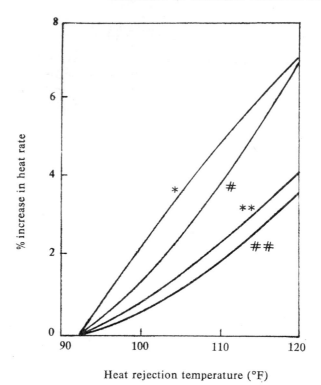

Figure 12.1 Sensitivity of various power conversion cycles to heat rejection temperature.

    * = 1000 psi, 544F Rankine Cycle (Typical of PWR or BWR)
    # = 1500 psi, 1450F Brayton Cycle (*i.e.*, Joule's cycle or gas-turbine cycle)
  ** = 2400 psi, 900/900 Rankine Cycle (Typical LMFBR) or HTGR (see next
        section)
  ## = 3600 psi, 1000/1000 Rankine Cycle (Typical Fossil-Fueled Plant)

c)   It has been found that Juvenile king salmon had a maximum temperature tolerance of 83°F for a 10-minute test and withstood instantaneous temperature rises of 25°F with no mortalities within 24 hours after testing. Both salmon and striped bass passed successfully the condenser tests, consisted of fish passed through an operating condenser at full plant load and with a 16°F temperature rise over a 3 to 5 minute period. These fish were observed for 10 to 21 days afterwards. In fact, juvenile king salmon did not suffer any mortalities until the shock temperature increase was 30°F or greater.

d)   The dissolved oxygen levels were not decreased in passing through the cooling water system, but the water merely became super-saturated with dissolved oxygen, according to actual measurements. The evidence indicates that the saturation levels of dissolved oxygen held in the heated water discharge drop only as fast as the temperatures decrease.

e)   Surveys showed that warm water discharged at the stream's edge "floated" out into the main channel and was then mixed quite rapidly with stream flow.

f)   Larger power plants will influence a lesser area per megawatt than would smaller power plants of an equivalent total output.

g)   Cooling towers increase the operational costs by about 0.2 to 0.4 mills/kwh (1 mill $= 10^{-3}$ ¢) over the life of a projected facility, in comparison to the fixed charge of 4.07 mills/kwh for the capital cost of a 1100 Mwe nuclear plant with capital estimated at 14% and 7000 hrs/yr.

A widely publicized case between power plant officials and conservationists on thermal pollution involves the Florida Power and Light Company and Biscayne Bay. The power generating complex proposed by the Company is such that at the maximum coolant intake rate of 4.5 million gallons of water per minute, the atomic units will gulp in, and discharge out, the equivalent volume of the entire Bay every month. It is estimated by the ecologists that this would raise the temperature of the water 10 to 18°F above its normal 92°F. The question is: What effect would it have on the aquatic life cycle and what should be done if the units are not allowed to be operated, thus failing to meet the expected power demand? The cooling towers alternative is ruled out because of cost, hurricane danger, and salt spray. Long pipeline and cooling ponds are also ruled out because of cost. The construction of a second nuclear plant at Crystal River was cancelled by the Florida Power Corporation and fossil-fueled plants were planned. Delays in construction and cost increases were cited as the major reasons.

C.   High Temperature Gas-cooled Reactor (HTGR)

The light water PWR (Westinghouse, Babcock and Wilcox, and Combustion Engineering) and BWR (General Electric) have been dominating the U.S. market and are becoming increasingly popular oversea. The most significant progress in the field of nuclear engineering is the recent entrance of the high temperature gas cooled reactor into the U.S. commerical market in the 1100 Mwe range. In this section, the HTGR (Gulf General Atomic) will be discussed.

P

1. Gas Cooling

The HTGR is helium cooled, unlike some of the European reactors cooled by carbon dioxide. Helium gas cooling has certain characteristics:

a) Single phase operation, without local boiling or voiding.

b) It can be insulated from the walls of its enclosure, making possible the use of the prestressed concrete pressure vessel (PCRV).

c) Low neutron absorption and moderation, enhancing the core neutron economy.

d) Chemical inertness and nonradioactive, making it easier to access the core for service. Since helium is optically transparent, in-core inspection is easier and infrared scanning may be used for measurement and control. Unlike the light water reactor, a gas cooled reactor has no N-16 gammas to be worried about. Because of the inertness, the designers can avoid the necessity for an intermediate loop for primary coolant isolation, as in the case for the liquid metal fast breeder reactor (below).

e) The steam generators and helium recirculators of a HTGR are inside the PCRV. The LOCA appears to be a much less severe accident for the HTGR as is for the LWR.

2. General Description

Certain important plant characteristics are listed below:

Reactor power, Mwt	3000
Net plant output, Mwe	1160
Primary coolant loops	6
Helium coolant temperature, in/out, °F	625/1413
Coolant pressure, psia	700
Throttle steam conditions, psig/°F	2400/950
Degree of superheat, °F	288
Final feed water temperature, °F	370
Reheat steam, psig/°F	575/1000
Auxiliary core cooling loops	3
Fuel type	carbide particle
Control rods	top mounted
Moderator and reflector	graphite
Reactor vessel	PCRV
Secondary containment	yes
Steam generator	once through integral reheat

Inherent with the high temperature coolant is the fact that a HTGR may reject about 25% less heat than LWR's of equivalent core thermal output.

If river or ocean once-through raw cooling water (RCW) is used and a 15°F rise is assumed for both a HTGR and a LWR, then the HTGR would use up less water and this is reflected in smaller circulating water pumps, lower intake velocities, and reduced environmental impact. Wet cooling towers or cooling ponds, if used, can similarly be smaller.

The enclosed volume of a PCRV is around 100,000 cubic feet, with outside cylindrical dimensions at about 90 feet high and 96 feet diameter. Due to the low heat capacity of the gaseous coolant, the secondary containment is only about half the size of a PWR dry containment with the same power output. An acceptable secondary containment may have 1.7 million cubic feet of free volume and design pressure at 33 psig.

Because of the high core temperature, an important fraction of the core neutrons has energies in or near the broad low lying resonance peaks occurring in the U-233 fission and capture cross-section. A very small shift in the average neutron energy arising from changes in the moderator temperature may cause steep changes in neutron utilization. At high burnups, the resonance structure of the fission product absorption spectra also became significant. The control and design of the HTGR must take these factors into consideration. Provision must also be given to accommodate the graphite dimensional shrinkage when it is irradiated at high temperatures. However, no special provisions are required in HTGR's to remove the stored energy (often called Wigner energy), which can accumulate in graphite when it is irradiated at low temperature, since the operating temperature is sufficiently high to continuously anneal the radiation induced lattice defects in the crystal structure of the graphite.

The prospect of a gas turbine to go with the HTGR (*i.e.*, without steam generator) is the subject of many recent discussions.

The HTGR is a low inventory, potential "near breeding" advanced converter. The basic fuel cycle used is the uranium/thorium. However, other fuel cycles can also be used. The availability of U-233 for the HTGR greatly enhance the fuel economics. A high gain breeder, such as the gas cooled fast breeder reactor (GCFR), may one day become part of a package in nuclear power technology.

D. Liquid Metal Fast Breeder Reactor (LMFBR)

From the data given in Table 3-2, it is seen that Pu-239 is more suitable for fast neutron fission and U-233 is more suitable for slow neutron fission. Radiative capture of neutrons in U-238 and Th-232 leads to Pu-239 and

U-233, respectively, through the following reactions:

$$U^{238} + n \rightarrow U^{239} \rightarrow Np^{239} + e$$
$$\phantom{U^{238} + n \rightarrow U^{239} \rightarrow Np^{239}} \hookrightarrow Pu^{239} + e$$

$$Th^{232} + n \rightarrow Th^{233} \rightarrow Pa^{233} + e$$
$$\phantom{Th^{232} + n \rightarrow Th^{233} \rightarrow Pa^{233}} \hookrightarrow U^{233} + e$$

U-238 has a relatively large fission cross-section only with neutrons of energy above 1 Mev, with several fast neutrons emitted for each fast neutron fission. The resulting Pu-239 is a useful fissil fuel. In order to sustain a fast chain reaction, the neutrons must not be slowed down by moderator. It is with the intention of making use of the U-238 and keeping the neutrons in the core at a fast spectrum to breed more fuel that liquid metal is selected as the coolant for the LMFBR. Whereas LWR's recover less than 2% of the energy locked in uranium reserves, the breeder will recover over 70%.

There are two major designs for LMFBR, the so-called "loop" type or "pot" type. Liquid sodium is the primary coolant for both types and both types have an intermediate heat exchanger (IHX), with non-radioactive sodium as secondary coolant.

In the "pot" system, the core, pump, and IHX are all positioned in a large tank of sodium. The secondary sodium flows from the IHX in the pot to the steam generator located outside of the tank. The most important design basis is that of assuring coolant boundary integrity, since the possibility of leaks and ruptures is minimized.

In the "loop" system, which is also sometimes called "pipe" system, the primary coolant is pumped through pipes to the intermediate heat exchangers located in vaults external to the reactor. The major design consideration for the pipe system is to assure core submergence by using elevated loops and guard tanks.

For the LMFBR, the primary system is a low pressure system, less than 250 psig, since the vapor pressure of liquid sodium is small in comparison to that of steam. The advantage of a low pressure primary system is obvious.

In a PWR, the reactor coolant flow is maintained at a constant value. The temperature drop across the core changes with load and is of the order of 60°F. The time rate of change of vessel temperature with load change is acceptably small. For a LMFBR, on the other hand, a constant primary coolant flow results in a temperature gradient of about 275°F across the vessel. The resulting thermal stress due to load change is a severe problem. This serves to illustrate some of the problems involved.

There are other variations to the pot system, the tank of which may be as large as 60 feet in diameter for a 1000 Mwt reactor. One of the possible designs can be classified as a two-pool system, with two control surfaces.

The U.S. has recently committed to building a LMFBR demonstration plant. It will be built near Oak Ridge, Tennessee. Involved organizations are the Commonwealth Edison Company and the Tennessee Valley Authority (TVA), forming the Project Management Corporation, and the AEC. Vendor or vendors to supply the nuclear steam supply system (NSSS) have not yet been decided when this was written. The power level will be between 300 and 600 Mwe.

E. Controlled Fusion Power

Fission energy is now more economical than fossil energy for large power plants. It is also much less polluting than conventional fossil plants. However, fission has certain drawbacks:

1. Due to low enrichment, bomb explosion is not possible, but meltdown and power excursions are credible accidents.

2. Fission products are radioactive. Shielding and radwaste storage are inherent problems.

3. The plutonium by-product is basically a weapon material. Large quantities of this may become a security problem to the industry and to world politics.

4. The fuel reserves are probably sufficient for the next century, but not inexhaustible.

It is because of these and other drawbacks that fusion power has been considered seriously. Since deuterium is in abundance in ordinary water, the fuel reserve is practically infinite. Fusion, by definition, has no fission product (but the tritium in the fusion cycle is a beta emitter). The most decisive advantage of a fusion reactor over a fission reactor is due to the fact that a runaway fusion power plant is not possible, making it likely that a future fusion plant may be directly inside (or under) a highly populated area. On the other hand, fusion reactors are not without problems. Relatively large neutron activation is just one of them.

To have a successful fusion reaction, the product of the plasma density and the confinement time must exceed a minimum. This minimum is at around 10^{14} ion-cm^{-3}-sec for an energy breakeven point (*i.e.*, input energy in heating the plasma = fusion energy output). The achievement of this has been called feasibility. The requisite ion temperature is at about 10 kev

(= 114 million degree Kelvin) for D-T reaction and at about 50 kev for D-D reaction.

Fusion reaction is being investigated by many countries. By far the most talkabout system is the Tokamak, devised by Artsimovich of Russia. It is a large toroidal system with thick copper walls. A toroidal winding encircles a magnetic core. Induced current is used to heat the plasma and a magnetic field generated from the toroidal winding contains the plasma. Stability has always been a problem in fusion study, but the Tokamak achieved for the first time an artificial thermonuclear temperature in 1969, even though it was only for about 20 milliseconds. An ion temperature of 0.4 kev was reached, below the energy breakeven point and not sufficient for a reactor.

Many other ideas are of interest. These include changing the winding geometries to achieve "pinch" effect for stability, metal "mirror" for radiation reflection, laser heating, etc.

Fusion reactor is not yet available. Scientists and engineers are determined to make the breakthrough. The implicit assumption is that the problem can be solved.

Chapter 13

CRITERIA, STANDARDS AND GUIDES

IN THE United States, there are many professional organizations and regulatory agencies, both national and statewise, that set requirements on the standards for nuclear reactors. Examples are the American Society for Mechanical Engineers (ASME), the American Nuclear Society (ANS), the Institute of Electrical and Electronics Engineers (IEEE), the Atomic Energy Commission (AEC), the Environmental Protection Agency (EPA), etc. In this chapter, some of the most important criteria, standards, and guides are presented. Since this text is not intended to be a handbook, the presentation is by no means complete and is only an attempt to indicate the scope.

A. Definitions and Explanations

Nuclear power unit. A nuclear power unit means a nuclear power reactor and associated equipment necessary for electrical power generation and includes those structures, systems, and components required to provide reasonable assurance the facility can be operated without undue risk to the health and safety of the public.

Loss of coolant accidents. Loss of coolant accidents mean those postulated accidents that result from the loss of reactor coolant at a rate in excess of the capability of the reactor coolant makeup system from breaks in the reactor coolant pressure boundary, up to and including a break equivalent in size to the double-ended rupture of the largest pipe of the reactor coolant system.

Single failure. A single failure means an occurrence which results in the loss of capability of a component to perform its intended safety functions. Multiple failures resulting from a single occurrence are considered to be a single failure. Fluid and electrical systems are considered to be designed against an assumed single failure if neither (1) a single failure of any active component (assuming passive components function properly) nor (2) a single failure of a passive component (assuming active components function properly), results in a loss of the capability of the system to perform its safety functions.

Anticipated operational occurrences. Anticipated operational occurrences mean those conditions of normal operation which are expected to occur one or more times during the life of the nuclear power unit and include but are not limited to loss of power to all recirculation pumps, tripping of the turbine generator set, isolation of the main condenser, and loss of all offsite power. *10 CFR.* The Code of Federal Regulations (CFR), title 10, is on Atomic Energy, Chapter 1 of which is on Atomic Energy Commission. Often, 10CFR20 is used. Part 20 is about Standards for Protection Against Radiation. Similarly, 10CFR50 is about Licensing of Production and Utilization Facilities and 10CFR100 is on Reactor Site Criteria.

B. General Design Criteria (GDC)

The Appendix A of 10CFR50 sets forth the General Design Criteria for Nuclear Power Plants. There are 55 of these, published on February 20, 1971, effective 90 days after publication. These criteria are listed in Table 13.1.

TABLE 13.1   General design criteria for nuclear power plants

Group	Title	Criterion No.
I.	Overall requirements:	
	Quality Standards and Records	1
	Design Bases for Protection Against Natural Phenomena	2
	Fire Protection	3
	Environmental and Missile Design Bases	4
	Sharing of Structures, Systems, and Components	5
II.	Protection by multiple fission product barriers:	
	Reactor Design	10
	Reactor Inherent Protection	11
	Suppression of Reactor Power Oscillations	12
	Instrumentation and Control	13
	Reactor Coolant Pressure Boundary	14
	Reactor Coolant System Design	15
	Containment Design	16
	Electrical Power Systems	17
	Inspection and Testing of Electrical Power Systems	18
	Control Room	19
III.	Protection and reactivity control systems:	
	Protection System Functions	20
	Protection System Reliability and Testability	21
	Protection System Independence	22
	Protection System Failure Modes	23
	Separation of Protection and Control Systems	24
	Protection System Requirements for Reactivity Control Malfunctions	25

Table 13.1 (contd.)

	Reactivity Control System Redundancy and Capability	26
	Combined Reactivity Control Systems Capability	27
	Reactivity Limits	28
	Protection Against Anticipated Operational Occurrences	29
IV.	Fluid systems:	
	Quality of Reactor Coolant Pressure Boundary	30
	Fracture Prevention of Reactor Coolant Pressure Boundary	31
	Inspection of Reactor Coolant Pressure Boundary	32
	Reactor Coolant Makeup	33
	Residual Heat Removal	34
	Emergency Core Cooling	35
	Inspection of Emergency Core Cooling System	36
	Testing of Emergency Core Cooling System	37
	Containment Heat Removal	38
	Inspection of Containment Heat Removal System	39
	Testing of Containment Heat Removal System	40
	Containment Atmosphere Cleanup	41
	Inspection of Containment Atmosphere Cleanup Systems	42
	Testing of Containment Atmosphere Cleanup Systems	43
	Cooling Water	44
	Inspection of Cooling Water System	45
	Testing of Cooling Water System	46
V.	Reactor containment:	
	Containment Design Basis	50
	Fracture Prevention of Containment Pressure Boundary	51
	Capability for Containment Leakage Rate Testing	52
	Provisions for Containment Inspection and Testing	53
	Systems Penetrating Containment	54
	Reactor Coolant Pressure Boundary Penetrating Containment	55
	Primary Containment Isolation	56
	Closed Systems Isolation Valves	57
VI.	Fuel and radioactivity control:	
	Control of Releases of Radioactive Materials to the Environment	60
	Fuel Storage and Handling and Radioactivity Control	61
	Prevention of Criticality in Fuel Storage and Handling	62
	Monitoring Fuel and Waste Storage	63
	Monitoring Radioactivity Releases	64

C. Safety Guides

The AEC publishes safety guides for water-cooled nuclear power plants and adds new ones occasionally. Table 13.2 lists the title of 29 guides available as of June 7, 1972. Table 13.3 lists the guides under development. The primary

TABLE 13.2   Safety guides for water-cooled nuclear power plants

Safety guide number	Title
1	Net Positive Suction Head for Emergency Core Cooling and Containment Heat Removal System Pumps
2	Thermal Shock to Reactor Pressure Vessels
3	Assumptions Used for Evaluating the Potential Radiological Consequences of a Loss of Coolant Accident for Boiling Water Reactors
4	Assumptions Used for Evaluating the Potential Radiological Consequences of a Loss of Coolant Accident for Pressurized Water Reactors
5	Assumptions Used for Evaluating the Potential Radiological Consequences of a Steam Line Break Accident for Boiling Water Reactors
6	Independence Between Redundant Standby (Onsite) Power Sources and Between Their Distribution Systems
7	Control of Combustible Gas Concentrations in Containment Following a Loss of Coolant Accident
8	Personnel Selection and Training
9	Selection of Diesel Generator Set Capacity for Standby Power Supplies
10	Mechanical (Cadweld) Splices in Reinforcing Bars of Concrete Containments
11	Instrument Lines Penetrating Primary Reactor Containment
12	Instrumentation for Earthquakes
13	Fuel Storage Facility Design Basis
14	Reactor Coolant Pump Flywheel Integrity
15	Testing of Reinforcing Bars for Concrete Structures
16	Reporting of Operating Information
17	Protection Against Industrial Sabotage
18	Structural Acceptance Test for Concrete Primary Reactor Containments
19	Nondestructive Examination of Primary Containment Liners
20	Vibration Measurements on Reactor Internals
21	Measuring and Reporting of Effluents from Nuclear Power Plants
22	Periodic Testing of Protection System Actuation Functions
23	Onsite Meteorological Programs
24	Assumptions Used for Evaluating the Potential Radiological Consequences of a Pressurized Water Reactor Gas Storage Tank Failure
25	Assumptions Used for Evaluating the Potential Radiological Consequences of a Fuel Handling Accident in the Fuel Handling and Storage Facility for Boiling and Pressurized Water Reactors
26	Quality Group Classifications and Standards
27	Ultimate Heat Sink
28	Quality Assurance Program Requirements (Design and Construction)
29	Seismic Design Classification

TABLE 13.3   Safety guides under development

1	Reactor Coolant Pressure Boundary Leakage Detection
2	Design Phase Quality Assurance Requirements
3	Quality Assurance Program Requirements (Operation)
4	Monitoring and Reporting of Environmental Levels
5	Diesel Generator Protective Interlocks
6	Use of IEEE Std 308-1971, "IEEE Standard Criteria for Class IE Electric Systems for Nuclear Power Generating Stations"
7	Availability of Emergency Power Sources
8	Physical Independence of Safety Related Electric Systems
9	Indication of Bypasses in Protection Systems and Engineered Safety Feature Systems
10	Post-Accident and Incident Monitoring
11	Flood Design Bases
12	Inservice Surveillance of Ungrouted Prestressing Tendons
13	Stainless Steel Weld Metal Microfissure Control
14	Design Loading Combinations for Fluid System Components
15	Design Loading Combinations for Primary Metal Containment Systems
16	Isolating Low Pressure Systems Connected to the Reactor Coolant Pressure Boundary and that Penetrate Primary Reactor Containment
17	Control of Lifting Equipment at Nuclear Power Plant Sites

purpose of the safety guides is to "describe and make available to the industry these methods of implementing the regulations and solutions to other safety issues that are acceptable to the regulatory staff and the Advisory Committee on Reactor Safeguards." Safety guides are not intended as substitutes for regulations. Compliance with safety guides is not required, but highly advisable.

D.   Codes

Certain subsections in 10CFR are of particular interest. These are described briefly below.

10CFR20. This is on Standards for Protection Against Radiation. It describes the control of releases of radioactivity to the environment and the guides for acceptable internal and external dose rates on-site and off-site, during normal operation. The protection criteria given in Chapter 9 of this text is very much parallel to that in 10CFR20.

10CFR50. This is on Licensing of Production and Utilization Facilities. It provides rules and regulations for licensing. Within 10CFR50, there are many appendices. For example,

Appendix A: On general design criteria, described earlier.

Appendix B: On quality assurance criteria.

Appendix C: On financial qualifications.

Appendix D: On specified environmental considerations.

Appendix I: On numerical guides for design objectives and technical specification requirements for limiting conditions for operation. The "as low as practicable" guidance on offsite dose is very demanding on the utility industry.

Appendix J: On test requirements for containment leak tightness, including Type A for integrated test, Type B for penetration test, and Type C for isolation valve test.

10CFR100.   This is on Reactor Site Criteria. It defines site evaluation factors and specifies acceptable off-site doses in the event of an accident.

E.   Emergency Core Cooling System—Interim Policy Statement

Protection against a highly unlikely LOCA has long been an essential part of the defense-in-depth concept used by the nuclear power industry and the AEC. A recent semiscale experiment on the blowdown system at the National Reactor Testing Station in Idaho appeared to show deviations from the predictions of the codes then in use. The emergency core cooling water was blocked and ejected from the system during blowdown. Although the test is not to scale, hearings are conducted on the ECCS and AEC has published the interim policy statement. (Federal Register, volume 36, No. 125, June 29, 1971). It says that on the basis of that day's knowledge, the performance of the ECCS is judged to be acceptable if the *calculated* course of the LOCA is limited as follows:

1.   The calculated maximum fuel element cladding temperature does not exceed 2300°F. This limit has been chosen on the basis of available data on embrittlement and possible subsequent shattering of the cladding. The results of further detailed experiments could be the basis for future revision of this limit.

2.   The amount of fuel element cladding that reacts chemically with water or steam does not exceed 1 percent of the toal amount of cladding into the reactor.

3.   The clad temperature transient is terminated at a time when the core geometry is still amenable to cooling, and before the cladding is so embrittled as to fail during or after quenching.

4.   The core temperature is reduced and decay heat is removed for an extended period of time, as required by the long-lived radioactivity remaining in the core.

F. ANS Safety Classes

Proposed ANS Safety Classes for PWR's, as given in the March 6, 1972, Draft are:

Safety Class 1

Safety Class 1, SC-1, applies to components whose failure could cause a Condition III or Condition IV loss of reactor coolant.

Safety Class 2

Safety Class 2, SC-2, applies to reactor containment and to those components:
a) Of the reactor coolant system not in Safety Class 1,
b) That are necessary to:
    remove directly residual heat from the reactor,
    circulate reactor coolant for any safety system purpose,
    control within the reactor containment radioactivity released,
    control hydrogen in the reactor containment, or
c) Of safety systems located inside the reactor containment.

Safety Class 3

Safety Class 3, SC-3, applies to those components not in Safety Class 1 or Safety Class 2:
a) The failure of which would result in release to the environment of radioactive gases normally required to be held for decay, or
    that are necessary to:
b) Provide or support any safety system function
c) Control outside the reactor containment air-borne radioactivity released, or
d) Remove decay heat from spent fuel.

G. Safety Analysis Report (SAR) Format

The regulatory staff of the USAEC issued in February, 1972, a standard format and content for the preliminary safety analysis report (PSAR) and final safety analysis report (FSAR). The titles of the 17 chapters are:
1. Introduction and General Description of Plant
2. Site Characteristics
3. Design Criteria—Structures, Components, Equipment, and Systems
4. Reactor

5. Reactor Coolant System
6. Engineered Safety Features
7. Instrumentation and Controls
8. Electric Power
9. Auxiliary Systems
10. Steam and Power Conversion System
11. Radioactive Waste Management
12. Radiation Protection
13. Conduct of Operation
14. Initial Tests and Operation
15. Accident Analysis
16. Technical Specifications
17. Quality Assurance

H. ASME Section III

The 1971 Edition of ASME Section III, "Nuclear Power Plant Components", presents the rules for construction of four classes (Class 1, 2, 3, and 4). Section III includes the requirements for piping, valves, and pumps, in addition to pressure vessels, which were given in the Pump and Valve Code and ANSI B31.7, "Nuclear Power Piping", prior to the issue of the 1971 edition of Section III. Section III does not classify an item but tells how to fabricate it once it has been classified. The code class of each nuclear component or system is determined by the owner or his agent.

I. Electrical Safety Class

Electrical systems and components are classified in accordance with "IEEE Criteria for Class IE Electric Systems for Nuclear Power Generating Stations", IEEE No. 308. There are three classes: I, II, and III. Another standard often referred to is "Criteria for Protection Systems for Nuclear Power Generating Stations", IEEE Std 279.

MISCELLANEOUS CONSTANTS

$\pi = 3.1415926535+$

base of exponential e $= 2.7182818284+$

$$e^{10} = 22026$$

$g = 32.172$ ft/sec^2 $= 980.665$ cm/sec^2 at sea level, lat. 45°

Avogadro's number $= 0.60247 \times 10^{24}$ mole^{-1}

Boltzmann's constant $k = 1.38044 \times 10^{-16}$ erg/°K

Planck's constant $h = 6.6254 \times 10^{-27}$ erg-sec

Stefan constant for black body $= 5.6686 \times 10^{-5}$ erg/cm^2-sec-(°K)^4

universal gas constant $R = 8.3166 \times 10^7$ erg/°K-mole

$$= 1.9869 \text{ cal } (15°\text{C})/°\text{K-mole}$$

$$\cong 53.36 \frac{(\text{lb/ft}^2)(\text{ft}^3)}{\text{degree Rankin}} \text{ for air}$$

$$\cong 85.6 \frac{(\text{lb/ft}^2)(\text{ft}^3)}{\text{degree Rankin}} \text{ for water vapor}$$

mass of hydrogen atom $= 1.67339 \times 10^{-24}$ g

velocity of sound in dry air at 0°C and 1 atm $= 1087.1$ ft/sec.

in water at 20°C $= 4823$ ft/sec

in pine wood $= 10,900$ ft/sec

in soft steel $= 16,410$ ft/sec

density of dry air at 0°C and 760 mm Hg $= 0.001293$ g/c.c.

heat of fusion of water at 0°C $= 79.63$ cal/g

heat of vaporization of water at 100°C $= 539.55$ cal/g

specific heat of air at constant pressure $= 0.238$

viscosity of water at 20°C $= 1.002$ centipoise

CONVERSION TABLE

1. *Length*

1 meter = 3.280833 feet

2. *Area*

1 barn = 10^{-24} cm^2

3. *Volume*

1 U.S. liquid gallon = 0.13368 ft^3 = 3.7854 liters
1 U.S. gallon of water at 62°F = 8.337 lbs = 3.7820 kg
1 British gallon = 0.16054 ft^3 = 4.5461 liters
1 British gallon of water at 62°F = 10 lbs

4. *Weight*

1 lb = 453.5924 g
1 short ton = 2000 lbs
1 long ton = 2240 lbs
1 metric ton = 1 tonne = 2204.62 lbs = 1000 kg
1 electron rest mass = 9.1083 × 10^{-28} g = 0.510976 Mev
1 neutron rest mass = 1.67470 × 10^{-24} g
1 U-235 rest mass = 235.11704 amu (physical scale)
1 atomic mass unit (amu) = 1.6598 × 10^{-24} g = 931.141 Mev

5. *Density*

1 g/c.c. = 62.43 lb/ft^3
density of air at STP = 0.001293 g/c.c.

6. *Force*

1 lb.wt. = 453.59 g.wt.
1 dyne = 2.2481 × 10^{-6} lb.wt.

7. *Pressure*

1 atm = 1.013246 bars = 1.013246 × 10^6 dynes/cm^2
    = 14.696 psi
    = 760 mm Hg (0°C) = 33.903 ft water (0°C)
 1 psi = 70.307 g/cm^2

Q

8. *Mass Flow Rate*

1 lb/hr = 0.126 g/sec
1 gpm = 0.002228 ft^3/sec = 0.06308 liter/sec

9. *Work and Energy*

1 BTU = 1054.8 watt-sec = 3.9292 × 10^{-4} horsepower-hour
1 cal = 4.186 watt-sec
1 g = 5.6100 × 10^{26} Mev
1 ev = 1.60206 × 10^{-19} watt-sec
1 watt-sec = 1 joule = 10^7 erg
1 kw-hr = 4.2 × 10^{-5} gram of U-235 in fission
        = 6.4 × 10^{-6} gram of H^3 in H^3(d, *n*)He^4 reaction
        = 3415 BTU
        = 2.66 × 10^6 ft-lb
        = 8.60 × 10^5 cal
        = 2.24 × 10^{19} Mev
        = 0.74 lb of coal (based on highest modern fossil power station
          efficiency of 12,500 BTU/lb-coal)

10. *Power*

1 kw = 3415 BTU/hr
     = 1.341 horse-power
     = 239 cal/sec

11. *Temperature*

0°C = 273°K = 32°F = 492°R (R = Rankine)
Δ1°C = Λ1°K = Δ1.8°F = Δ1.8°R (R = Rankine)

12. *Thermal Conductivity*

1 BTU/hr-ft-(°F/ft) = 0.0173 watt/cm^2-(°C/cm)
                    = 4.13 × 10^{-3} cal/sec-cm^2-(°C/cm)

13. *Heat Capacity*

1 BTU/lb-°F = 1 cal/g-°C

14. *Heat Flux*

1 BTU/hr-ft^2 = 3.15 × 10^{-4} watt/cm^2
              = 7.53 × 10^{-5} cal/sec-cm^2

15. *Heat Transfer Coefficient*

1 BTU/hr-ft^2-°F = 5.67 × 10^{-4} watt/cm^2-°C
                  = 1.36 × 10^{-4} cal/sec-cm^2-°C

16. *Radiation*

1 curie $= 3.7 \times 10^{10}$ disintegrations/sec
1 roentgen $= 1$ esu/c.c. std air
$\quad\quad\quad\quad\;\; = 2.083 \times 10^9$ ion pairs/c.c. std air
$\quad\quad\quad\quad\;\; = 1.61 \times 10^{12}$ ion pairs/g std air
$\quad\quad\quad\quad\;\; = 6.77 \times 10^4$ Mev/c.c. std air
$\quad\quad\quad\quad\;\; = 83.8$ erg/g air
$\quad\quad\quad\quad\;\; = 5.24 \times 10^7$ Mev/g air
1 rep $= 93$ ergs/g tissue
1 rem $= 93/$RBE erg/g tissue
1 rad $= 100$ ergs/g of any material

GREEK ALPHABET

Greek letter	Greek name	English equivalent	Greek letter	Greek name	English equivalent
A α	Alpha	a	N ν	Nu	n
B β	Beta	b	Ξ ξ	Xi	x
Γ γ	Gamma	g	O o	Omicron	ŏ
Δ δ	Delta	d	Π π	Pi	p
E ε	Epsilon	ĕ	P ρ	Rho	r
Z ζ	Zeta	z	Σ σ	Sigma	s
H η	Eta	ē	T τ	Tau	t
Θ θ ϑ	Theta	th	T υ	Upsilon	u
I ι	Iota	i	Φ φ φ	Phi	ph
K κ	Kappa	k	X χ	Chi	ch
Λ λ	Lambda	l	Ψ ψ	Psi	ps
M μ	Mu	m	Ω ω	Omega	ō

SUBJECT INDEX

Abrupt changes in flow area   121
Absolute deviation   149
Absorbed dose   134
Absorption   1
Absorption coefficient,
  energy   136
  mass   136
Absorption cross-section   12
Absorption cross-section,
  resonance   12,107
Absorption,
  resonance   59
Accident   45, 138
Activation   17
Activation,
  neutron   158
Activity   18, 134
Adu process   193
Age   63, 70, 90, 98
Age diffusion theory   45
Age equation   62
Aluminum   168
Ammonium diuranate process   193
Annihilation gamma   158
Atomic bomb   100
Atomic mass unit   7
Average kinetic energy   12
Average life of isotope   5
Average speed of molecules   12
Avogadro number   20, 24

Bare homogeneous core   73
Barn   12
Barn book   36
Becker, E. W.   136
Bessel equation   93
Bessel function   93
Bessel,
  modified equation   93
BF$_3$ counter   15, 148
binding energy   3
Biological variability   137
Boiling   119

Boiling,
  nucleate   120
  partial film   120
Boiling water reactor   131, 194
Boltzmann constant   11
Borda-Carnot expression   122
Boron   16, 139
Boron,
  cross-section   28
Boron trifluoride   16
Borst-Wheeler curve   35
Bragg's condition   17
Brayton cycle   208
Breeder   38, 131
Breeding   38, 131
Breit-Wigner formula   60
Bremsstrahlung gamma   158
Broadening,
  Doppler   107
Brown oxide   189
BSR   184
Btu   24
Bubble chamber   148
Buckling   62
Buckling,
  geometric   74
  material   75
Buildup factor   164
Bulk shielding reactor   184
Burnable poison   108
Burnout heat flux   120
Burnup   22, 107, 194
Bvu   113

Cadmium   16
Cadmium,
  cross-section   28
Californium   15
Calories   24
Capture gammas   158, 179
Capture,
  radiative   2
Cascade chain   193

Center of mass   68
Centipoise   113
Centrifuge gas process   191
Cerenkov detector   148
Cf-252   15
Chain reaction   23
Chamber,
    compensated ion   20
    fission   19
    pressure suppression   49, 53
Channeling effect   90
Charcoal adsorber   56
Chauvent's criterion   151
Chemical detector   19
Chemical dosimeter   148
Chop-leach method   196
Chopper,
    neutron   16
Cloud chamber   148
Co-60   153
Collided flux   164
Collisions,
    elastic and inelastic   57
Compensated ion chamber   20
Complete film boiling   120
Compound nucleus   1, 2
Compton scattering   160, 179
Concentrate purification   189
Concentration,
    maximum permissible   139, 141
Concrete   168, 169, 179
Conduction   111
Constants
    material   122
    thermal   122
Containment vessel   42, 53, 54, 55
Contraction,
    flow area   122
Control   97
Control,
    reactor   47, 97, 105
Control rod   43
Control rod,
    multiple   104
    off centered   104
    partially inserted   104
Control rod theory   102
Convection   111

Convection,
    single phase   119
Conversion,
    dose to flux   144
Converter   38, 131
Coolant   42, 43, 46
Coolant density   123
Coolant temperature distribution   115
Coolant viscosity   113, 127
Cooling,
    core standby system   50
    towers   209
Copper   21
Core   41, 47
Core,
    bare homogeneous   73
Core standby cooling system   50
Cost,
    power plant   199
Counter—see also detector
Counter,
    BF$_3$ 15, 148
    Cutie Pie   146
    dead time of   149
    fast neutron   16
    film dosimeter   148
    Geiger   145
    ionization   146
    long   16
    Ohmart   147
    pocket meter   146
    proportional   147
    proton recoil   19
    scintillation   16, 147
    statistics of   149
    other types of   148
Counting,
    statistics of   149
Critical heat flux   119, 120
Criticality condition   66, 73, 93
Cross-section   1
Cross-section,
    absorption   12
    boron capture   28
    cadmium   28
    differential   162
    fission   12, 27, 36
    indium   28

Cross-section—*cont.*
  Klein Nishina   162
  macroscopic   12
  microscopic   12
  radiative capture   12
  relaxation   168
  removal   168
  resonance absorption   12
  scattering   12
  thermal fission   27, 36
  thermal neutron   24, 30
  total   12, 28
  uranium 238   28
Crystal spectrometer,
  neutron   17
Curie   13, 134
Current density   77
Current density,
  neutron   77
Cutie Pie counter   146
Cycle,
  neutron   57
Cylindrical,
  Laplacian   73
Cylindrical,
  solid source   177

Damage to fuel   195
Dead time   149
Dead time of counter   149
DeBroglie wavelength   17
Decay constant   5
Decay gammas   34, 157
Decay power   36
Decay,
  constant   5
  radioactive   1, 4
  successive   5
Delayed neutrons   32, 102, 158
Delayed neutron fraction   34
Density,
  coolants   123
  steam and water   123
Depletion,
  fuel   108
Detection,
  fast neutron   148
Detection,

neutron   15, 148
Detection,
  radiation   133, 143
Detectors—see also counter, chamber
Detectors,
  chemical   19
  compensated ion chamber   20
  Fricke   19
  gas filled   145
  photographic emulsion   19, 148
  semiconductor   19, 148
  threshold   18
Deuterium   7
Deviation,
  absolute   149
  standard   149
Differential cross-section   162
Diffraction,
  neutron   17
Diffusion equation   61
Diffusion,
  gaseous   191
  neutron   60
  one group method   45, 73, 75
  two group method   80
Diffusion length   71, 90
Diffusion length,
  thermal   65
Disk source   173
Disposal,
  waste   196
Distributions,
  flux   73
  Gaussian   70, 150
  Maxwellian   11, 25
  normal   150
Dollar   99
DOP   56
Doppler broadening   107
Doppler coefficient   48
Dose to flux conversion   144
Dose,
  absorbed   134
  accidental   138
  accumulated   139
  effect of   138
  emergency   140
  fatal   138

internal   140
    maximum permissible   139
    medical   140
E-integral   173
Earthquake   53
Economics,
    fuel   187, 198, 199
Effective multiplication factor   66
Elastic collision   57
Electromagnetic separation   193
Emergency core cooling system   54, 220
Emergency dose   140
Emergency gas treatment system   51, 56
Emulsion,
    photographic   19, 148
Energy absorption coefficient   136
Energy determination,
    neutron   15
Energy,
    fission   23, 25
Energy of neutron   14
Engineered safety feature   45, 54
Enrichment   190
Environmental pollution   206
Epithermal neutrons   71
Equivalent hydraulic diameter   111, 113
Erg   24
Error function   150
Error,
    probable   151
    standard   151
Excess multiplication factor   98
Excess reactivity   106, 107
Excitation energy   3
Expansion,
    flow area   121
Explosive,
    nuclear   205
Exponential integral   173
Exposure   134
Exposure,
    accidental   138
    occasional   138
    see also dose
Extrapolation distance   74, 103

Fabrication of fuel   194
Facilities   41

Fanning equation   120
Fast breeder reactor   131, 211
Fast fission factor   64, 85, 90
Fast flux   82
Fast group   80
Fast neutron   179
Fast neutron counter   16
Fast neutron detector   148
Fast non-leakage probability   65
Fast removal   168, 179
Fatal dose   139
Fermi age   98
Fick's law   60
Film boiling,
    complete   120
    partial   120
Film dosimeter   148
F-integral   173
Fission   2
Fissionable materials   23
Fission chamber   19
Fission cross-section   27, 36
Fission energy   23
Fission fragment   2, 24
Fission fragment energy distribution   26
Fission fragment velocity distribution   26
Fission gamma,
        energy spectrum of prompt   24, 29
Fission neutron   15, 24, 25
Fission product poisoning   107
Fission prompt neutrons   158
Fission yield   26
Flash reactor   107
Fluid flow   120
Fluorination   190
Flux   12, 65
Flux to dose conversion   144
Flux,
    collided   164
    distributions   73
    fast   82
    thermal   82
    uncollided   164
Four factor formula   65
Fourier's equation   111
Fractional distillation   190
Fragment,
    fission   2, 24

Fragment,
  energy distribution of fission 26
  velocity distribution of fission 26
Freon 56
Fricke method and detector 19
Frictional pressure drop 120
Fuel 43, 47
Fuel cycle 187
Fuel depletion 108
Fuel economics 198
Fuel fabrication 194
Fuel loading 197
Fuel reprocessing 196
Fuel temperature 48, 117
Fuel temperature distribution 114
Fusion 2, 213

Gammas,
  capture 158, 179
  delayed 34, 157
  interaction with materials 158
  prompt 2, 34
  prompt fission 157
    energy spectrum 29
  types of 157
Gas centrifuge process 191
Gas cooled fast breeder reactor 209
Gas filled detectors 145
Gas turbine cycle 208
Gaseous diffusion 191
Gaussian distribution 70, 150
GCFBR 211
Geiger counter 145
General design criterion 216
Generation time 98
Genetic effect 137
Geology 52
Geometric buckling 74
Graetz number 113
graphite 168, 170
Gravitational acceleration constant 113

Half life 5
Heat capacity 111
Heat capacity,
  coolant 113, 125
Heat flux
Heat removal system 108, 111

Heat transfer coefficient 111, 112, 113
Heterogeneous 43
High efficiency particular filter 56
Heterogeneous cell 85, 114
HFIR 15
High flux isotope production reactor 15
High temperature gas cooled reactor 209
Homogeneous 43
Homogeneous,
  bare core 73
  reflected core 77
Horsepower-hour 24
HTGR 209
Hydraulic,
  equivalent diameter 111, 113
Hydrology 52

Ice condenser 51, 53
Indium,
  cross-section of 28
Inelastic collision 57
Inelastic scattering gamma 157
Infinite multiplication factor 65
Infinite plane source 173
Inhour 99
Inhour equation 98, 99, 102
Interaction of gamma with matter 158
Interim criteria 55, 220
Internal dose,
  accumulated 140
Ion chamber, 146
  compensated 20
Ion exchange method 189
Ionization 15
Iron 168, 169
Isolation 55

Joule's cycle 208

Kernel 171
Kilowatt-hour 24
Kinematic viscosity 113
Kinetic equation 97
Kinetics,
  reactor 97
Klein-Nishina cross-section 162

Laboratory system

Laminar flow   113, 121
Laplacian   73
Latent period   137
Lead   168, 169
Lead,
    mass attenuation coefficient   161
Leakage,
    neutron   60
Lethargy   59, 70
Light water cooled reactor   187
Linear energy transfer   134
Line source   172
Liquid drop model   3
Liquid metal cooled fast breeder
    reactor   211
LMFBR   211
Loading of fuel   197
Logarithmic energy decrement   59
Long counter   16
Loss of coolant accident (LOCA)   215

Macroscopic cross-section   12
Man,
    standard   133
Mass absorption coefficient   136
Mass attenuation of lead   161
Material buckling   75
Material damage   195
Material,
    density   123, 124
    fissionable   23
    heat capacity   125
    thermal conductivity   126
Maximum permissible
    concentration   139, 141
Maximum temperature   117
Maxwellian distribution   11, 25
Mean life of isotope   40
Megawatt thermal   129
Meteorology   52
Method of separation of variables   73
Metric ton   195
Mev   24
Microscopic cross-section   12
Migration length   71
Mining,
    fuel   187
Missile   52

Modified one group theory   95
Monoenergetic   13
Monte Carlo method   166
Most probable energy   12
Most probable speed   11
Multigroup theory   45
Multiple control rod   104
Multiplication factor,
    effective   66
    excess   98
    infinite   65
    prompt excess   98

NERVA   205
Neutrinos   2, 24
Neutron chopper   16
Neutron crystal spectrometer   17
Neutron current density   77
Neutron cycle   57
Neutron detection   15, 148
Neutron diffraction   17
Neutron diffusion   60
Neutron energy determination   15
Neutron leakage   60
Neutron,
    cross-section of thermal   24
    delayed   2, 32
    energies of   14
    fast counter for   16
    fast   16, 179
    fission   15, 24
    prompt   2
Neutron rest mass   17
Neutron slowing down   57
Neutron source   13, 158
Neutron,
    thermal   11, 15
Neutrons,
    types of   158
Newton's law of cooling   112
Non-leakage probability,
    fast   65
    thermal   65
Non-1/v factor   29, 36, 37
Normal distribution   150
Nozzle,
    separation method   192

Nu
Nuclear engineering   41
Nuclear engine for rocket vehicle application   205
Nuclear reactions   1
Nucleate boiling   120
Nusselt number   113

One delayed neutron group approximation   100
One group diffusion method   45, 73, 75
One group theory,
   modified   95
Off centered control rod   104
Ohmart cell   147
Orange oxide   190
Ore   187

Pu-239   152
Pair production   164, 179
Partially inserted rod   104
Pasquill classification   52
Peclet number   113
Period,
   latent to radiation   137
   reactor   99
Photoelectric effect   159, 179
Photographic emulsion   19, 148
Photomultiplier   147
Photoneutrons   13, 158, 179
Photoneutrons,
   angular distribution of   160
Planck constant   17, 159
Plane,
   disk source   173
   infinite source   173
Plant,
   reactor   51
Plowshare   205
Plutonium cycle   196
Pocket meter   146
Poise   113
Poison,
   burnable   108
Poisoning,
   fission product   5, 107
Point kernel   171
Pollution,
   air   206

thermal   207
Power density   131
Prandtl number   113
Pressure drop,
   frictional   120
Pressure suppression chamber   49, 53
Pressurized water reactor   128, 131, 186, 194
Prestressed concrete pressure vessel   210
Probability—see statistics
Probable error   151
Processing of fuel   187, 196
Prompt beta   34
Prompt critical   100
Prompt excess multiplication factor   98
Prompt fission gamma   29, 157
Prompt fission gamma energy spectrum   24, 29
Prompt gamma   2, 34
Prompt jump   102
Prompt neutrons   2
Proportional counter   147
Proton recoil counter   19
Pulse height analyzer   146
Purex method   196
Purification of ore   189

Q-energy   2
Quality factor   134

Rad   134
Radiation   35, 111
Radiation damage   195
Radiation detection   143
Radiation hazard   133
Radiation protection guide   139
Radiation source   157
Radiative capture   2
Radiative cross-section   12
Radioactive decay   1, 4
Radionuclides,
   unidentified   140
Radiosensitivity   137
Radiotoxicity   142
Random sampling method   166
Rankin cycle   208
Reaction,
   chain   23
   nuclear   1

Reactivity 98, 99, 104
Reactivity,
  excess 106, 107
Reactor 41, 45
Reactor,
  boiling water 131, 194
  bulk shielding 184
  Flash 107
  gas cooled 209
  nuclear rocket 131, 205
  pressurized water 128, 131, 186, 194
  see specific type
Reactor control 97, 105
Reactor characteristics 45
Reactor kinetics 97
Rectangular,
  Laplacian 73
Relative biological effectiveness 134, 137
Relaxation length 168, 179
Reflector 42, 43, 77
Rem 136
Removal cross-section 168, 179
Reprocessing of fuel 196
Resonance absorption and scattering 59
Resonance absorption cross-section 107
Resonance cross-section 12
Resonance escape probability 64, 85, 89
Reynolds number 113, 121
Rocket,
  nuclear reactor 131
Rod drop method 104
Rod,
  fully inserted 103
  multiple control 104
  off centered 104
  partially inserted 104
Rod worth 106
Roentgen 134
Root-mean-square speed 12
Russian Roulette 167

Safety class 221
Safety feature 45, 54
Safety guides 217
Scale of atomic mass 7
Scattering,
  Compton 160, 179
Scattering cross-section 1

Scattering—*cont.*
Scattering,
  elastic 1
  inelastic 1, 157
  resonance 59
Scintillation counter 147
Scram 100
Sec-integral 173
Secondary containment 54
Seismic 51, 53
Semiconductor detector 19
Semiempirical mass formula 3
Semi-infinite slab source 176
Separation nozzle method 192
Separation of variables,
  method 73
Separative work unit 198
Shielding 42, 157
Single failure 215
Site 52
Slowing down density 62, 70, 88
Slowing down length 71
Slowing down,
  neutron 57
Sm-149 107
SNAP 201
Solid cylindrical source 177
Solid sphere source 178
Solid state detector 148
Solvent extraction method 189
Sources,
  gamma 157
  neutron 158
  radiation 157
Space system 201
Spallation 2
Spark chamber 148
Spectrometer,
  neutron crystal 17, 148
Sphere,
  solid source 178
Spherical,
  Laplacian 73
Spherical surface source 175
Sr-90 153
Standard deviation 149
Standard error 151
Standard man 133

Statistics of counting   149
Steam,
   density   123, 124
   heat capacity   125
   thermal conductivity   126
Stefan's constant   112
Sterility   138
Subcritical   66
Successive decays   5
Supercritical   66
Suppression chamber   49, 53
Surface,
   spherical source   175
Systems for Nuclear Auxiliary Power   201

Temperature coefficient   106
Temperature distribution   114
Temperature distribution,
   coolant   115
   fuel   115
Temperature effect   105
Temperature,
   maximum   115
Time-of-flight method   16
Thermal conductivity   111, 113, 126
Thermal constants,
   material   122
Thermal diffusion length   66
Thermal fission cross-section   27, 36
Thermal flux   82
Thermal neutron   11, 15
Thermal neutron cross-section   24, 30
Thermal non-leakage probability   65
Thermal pollution   207
Thermal utilization   63, 85, 89
Thorium   23
Threshold detector   18
Threshold energy   2
Tokamak   214
Ton,
   metric   24, 195
Tonne   24, 195
Tornadoes   51, 52
Total cross-section   28
Transmutation   2, 4, 15

Transport theory   45
Turbulent flow   113, 121
Two group diffusion method   80
Two group theory   80
Two-region reactor   80

Uncollided flux   164
Unidentified radionuclides   140
Uranium   23, 170, 187
Uranium dioxide   189, 194
Uranium,
   fission cross-section of   28
Uranium haxafluoride   189, 190
Uranium oxide   187
Uranium resource   200
Uranium trioxide   190
Uranyl nitrate hexahydrate (UNH)   190

Vessel   42, 46
Viscosity   113, 127
Void coefficient   48
Volatilization   190
Volume expansion coefficient   105

Waste disposal   196
Water   168, 170
Water,
   density   123, 124
   heat capacity   125
   thermal conductivity   126
Water Quality Act   207
Watt-sec   24
Way and Wigner formula   35
Whole-body dose   139
Wigner and Way formula   35
Wind   51
Worth,
   control rod   106

Xe-135   107
Xe-135 poisoning   107

Yellow cake   187

Zirconium   194